INNOVATORS

Also by Donald Kirsch

The Drug Hunters: The Improbable Quest to Discover New Medicines (with Ogi Ogas)

INNOVATORS

*16 Visionary Scientists
and Their Struggle for Recognition—
from Galileo to Barbara McClintock
and Rachel Carson*

DONALD R. KIRSCH

Arcade Publishing • New York

First Edition

Arcade Publishing books may be purchased in bulk at special discounts for sales promotion, corporate gifts, fund-raising, or educational purposes. Special editions can also be created to specifications. For details, contact the Special Sales Department, Arcade Publishing, 307 West 36th Street, 11th Floor, New York, NY 10018 or arcade@skyhorsepublishing.com.

Arcade Publishing® is a registered trademark of Skyhorse Publishing, Inc.®, a Delaware corporation.

Visit our website at www.arcadepub.com.

10 9 8 7 6 5 4 3 2 1

Library of Congress Cataloging-in-Publication Data is available on file.
Library of Congress Control Number: 2023940311

Cover design by Erin Seaward-Hiatt
Cover photography: Jupiter image © Freak-Line-Community/Wikimedia Commons; iceberg © Michael Leggero/Getty Images; corn © Oliver Helbig/Getty Images

ISBN: 978-1-956763-39-3
Ebook ISBN: 978-1-956763-82-9

Printed in the United States of America

Contents

Introduction
Why Science Is the Slow Lane to Recognition and Fame

Everyone knows that great new things spring from the energy and innovations of youth. There are many examples of this in multiple domains. A young actor gives a terrific performance in a movie and instantly becomes a star, as John Wayne did in *Stagecoach* at the age of thirty-two. At the age of eleven, Anna Paquin won the 1994 Best Supporting Actress Oscar for her performance in *The Piano*, and Adrien Brody won the 2003 Best Actor Oscar at twenty-nine for his performance in *The Pianist*. In a different category, Damien Chazelle (my daughter's precocious classmate in elementary school) won the 2017 Best Director Oscar at the age of thirty-two for the movie *La La Land*.

Authors can become phenomenally successful at a young age too. Mary Shelley published her internationally acclaimed novel *Frankenstein* when she was twenty-one years old. When Helen Keller published her autobiography, *The Story of My Life*, she was twenty-two. Norman Mailer published his best-known novel, *The Naked and the Dead*, when he was just twenty-five. It was on the *New York Times* bestseller list for sixty-two weeks and is widely considered to be one of the best novels of the twentieth century.

Young musicians come up with a new musical style, their songs top the charts, and they instantly become pop stars. In 2008

Lady Gaga rose to stardom with the release of her album *The Fame*. She was twenty-two years old. At the age of twenty-eight, Artie Shaw recorded "Begin the Beguine," which launched his career as one of the top band leaders of the Big Band era. Frank Sinatra released his first big hit record, *Polka Dots and Moonbeams*, with the Tommy Dorsey Orchestra at the age of twenty-four and for the rest of his life basked in the adoration of a huge number of loyal and supportive fans. The Beatles became international rock stars in 1964. Ringo Starr and John Lennon were twenty-four years old, Paul McCartney was twenty-two, and George Harrison was twenty-one.

Business entrepreneurs, too, can hit it big at a young age. Bill Gates cofounded Microsoft in 1975 at the age of twenty with childhood friend Paul Allen. Allen was twenty-two. Microsoft would become the world's largest personal computer software company. At the age of twenty-one, Steve Jobs cofounded Apple in 1976 with twenty-six-year-old friend Steve Wozniak. Apple is currently the world's largest technology company by revenue. Google was founded in 1998 by Larry Page and Sergey Brin, two PhD students at Stanford University in California. Page and Brin were both twenty-five years old. Google has become one of the industry leaders in search engine technology, cloud computing, quantum computing, and artificial intelligence.

Facebook was founded in 2004 by Mark Zuckerberg with fellow Harvard College students and roommates Eduardo Saverin, Andrew McCollum, Dustin Moskovitz, and Chris Hughes. All of them were around twenty years old at the time. As of 2020, Facebook claimed 2.8 billion monthly active users and ranked seventh in global internet usage. Jeff Bezos founded Amazon from his garage in Bellevue, Washington in 1994, when he was thirty years old.

Certainly, success and recognition in these fields did not come instantaneously or without work. As a general rule, in addition to skill, talent, preparation, and a truly innovative approach to their endeavors, all of these people required two things to be successful. First, they were in the right place at the right time. Actors need the opportunity for a role that will showcase their talent to become available as well as an audience receptive to it. Writers and musicians need an environment able to accept their new prose or musical and performance styles and vision. Entrepreneurs need a marketplace ready to embrace their new product or the conditions for such a marketplace to arise. Second, all of these people must get their creation in front of that appreciative audience or in that marketplace: that is, they need the right break. When these things happen, the stars align for them.

Science is different. Recognition and fame can be agonizingly slow in science. In no small part, this is because scientists often must convince people that something is correct that defies common sense. The Earth is flying around the sun at 67,000 miles an hour, yet we do not feel we are moving. Heritable traits like eye and hair color seem to appear and disappear at random, despite the fact that these traits are controlled by clearly defined genetic laws. Germs are invisible to the naked eye, but they can easily and quickly kill a strong, healthy human.

In addition, scientists are by culture and training resistant to accepting anything new without overwhelming evidence. Scientists are professional skeptics, compelled to examine the evidence and question claims about it. One of the most damning scientific criticisms is to have your contention dismissed as only "a hand-waving argument"; that is, emphasized by some sort of a supportive gesture with your hands and saying "You know," but lacking substantiating data. The nineteenth-century chemist and microbiologist Louis Pasteur advised his students who were writing up their discoveries to "make it seem inevitable." Key opinion leaders in the discipline need to be convinced, and these leaders generally are attached to the prevailing older idea.

There of course are examples of scientists rising to the top of their profession quickly and at a young age, but such stories are notable for their rarity. I had a biology professor who wrote a scientific paper when he was a graduate student that became one of the most highly cited publications in the year it was published. This propelled him when he was in his late twenties to become an Ivy League professor who was soon granted tenure for life. After that, it mattered very little what he did in his later career. From the age of thirty onward, he remained a big deal as a scientist and professor at an Ivy League institution.

James Watson, who together with Francis Crick and Maurice Wilkins—and using data provided by Rosalind Franklin—determined the structure of DNA, is a well-known example of a young scientist whose accomplishments were quickly recognized. Watson was awarded the Nobel Prize for this discovery at the age of thirty-four, only nine years after the paper he published with Crick and Wilkins on DNA structure appeared in the journal *Nature*. Watson was significantly younger than the other two scientists with whom he shared the prize. Crick and Wilkins were both forty-six years old when they won it, still extremely young to be Nobel laureates.

One of my classmates in graduate school wrote his doctoral thesis on a major scientific breakthrough in quantum physics called asymptotic freedom, the theory that says that when subatomic particles called quarks come really close to one another they are no longer held together by the strong interaction. Thirty years later he won the Nobel Prize in Physics for this work, experimental studies that he had carried out when he was in his early twenties. This may sound like a long wait, but at the age of fifty-three he was one of the youngest scientists ever to achieve the most highly sought-after recognition in physics.

Such stories of success at a relatively young age are extremely unusual for a scientist. In science it commonly takes many, many years to win recognition. Most scientists achieve recognition and validation only very late in their careers and often only after surmounting numerous obstacles and enduring many disappointments. Ninety-two-year-old vocalist Tony Bennett stuck out like a sore thumb among the youthful musicians at the Grammy Awards presentation ceremony in 2018, but he would have fit right in on the stage of the Royal Swedish Academy of Sciences, receiving a Nobel medal in a scientific discipline. In the past decade, the average age of men and women awarded the Nobel Prize was seventy-one for chemistry, sixty-eight for physics, and sixty-eight for physiology or medicine.

Max Planck was an early twentieth-century physicist best known for showing how light carries energy. Planck determined that light was at the same time both a wave and a particle. This is impossible to visualize. To common sense, something cannot be both a particle and a wave but must be either one or the other. Planck's idea confounded common sense and, despite the fact that he was correct, he had an extremely hard time convincing other scientists that he was right. In his scientific autobiography Planck wrote:

A new scientific truth does not triumph by convincing its opponents and making them see the light, but rather because its opponents eventually die and a new generation grows up that is familiar with it. . . . An important scientific innovation rarely makes its way by gradually winning over and converting its opponents: it rarely happens that Saul becomes Paul. What does happen is that its opponents gradually die out, and that the growing generation is familiarized with the ideas from the beginning: another instance of the fact that the future lies with the youth.

Planck's idea about how scientific breakthroughs become accepted, known as Planck's principle, is often summarized as "Science progresses one funeral at a time." Scientists achieve recognition and fame for their new ideas only after the supporters of the older idea die out.

Discovering something new and of real importance in science takes years and years of experimental effort. Experimentation is slow; experiments often fail for technical reasons and have to be repeated to get all the conditions just right; and important new findings aren't based on a single result but rather on a large body of data that supports the new idea and eliminates a long list of possible alternate explanations. Once all that has happened, the real work begins of trying to convince the scientific establishment of the validity of what you have done.

Put simply, the scientific community doesn't want every crazy new idea to be quickly accepted as scientific truth. Innovative ideas must be vetted slowly and carefully before they can become accepted. The more wild and crazy an idea appears, even if it is correct, the more long-term scrutiny it undergoes. Commonly, new ways of thinking that dramatically move science forward are the ones that take the longest time to become accepted.

Medical researcher Jeremiah Stamler expressed Planck's idea in another way. It took a lifetime of effort for Stamler to convince the medical community that eating a healthy diet, low in sodium and cholesterol, and exercising and not smoking, would reduce the likelihood of heart disease and strokes. Prior to Stamler's work, almost all physicians believed that none of these things mattered for heart health.

One of Stamler's mentors, a top cardiology researcher named Louis Katz, almost convinced Stamler not to go into medical research. Toward the end of his life, Stamler was interviewed for a newspaper article and told the reporter that Katz had advised him, "Why the hell do you want to go into research? You never win. When you first discover something, people will say, 'I don't believe it.' Then you do more research and verify it and they'll say, 'Yes, but . . .' Then you do more research, verify it further, and they'll say, 'I knew it all the time.'" After many decades of medical research, Stamler came to realize that Katz was right.

I had a very minor brush with this in my own career. In the early 1980s I took my first job working in drug discovery for a large American pharmaceutical company. My assignment was to find new drugs to treat bacterial and fungal infections, but my boss also encouraged me to think of new ideas to treat diseases of all different types. For my principal assignment, I spent most of my time reading the scientific literature. But I also spent some time reading general research papers on the newest advancements outside the infectious disease field.

In the early 1980s, the oncogene theory was being developed by laboratories around the world to explain the cause of cancer. The theory says that the growth of cancer cells is caused by mutant genes called oncogenes. Research papers on the oncogene theory were appearing in many of the top scientific journals. At the time, drug treatment for cancer depended upon the cytotoxic chemotherapeutic agents. These drugs killed rapidly growing cells and were very toxic to cancer cells, which grow rapidly, but there are many rapidly growing healthy cells in the body too, and these rapidly growing healthy cells are also killed by the cytotoxic chemotherapeutic agents.

Killing the normal rapidly growing cells causes the terrible side effects produced by the cytotoxic chemotherapeutic agents: nausea and vomiting, hair loss, immune suppression, et cetera. It would be ideal to be able to kill cancer cells without damaging the healthy cells. It occurred to me that the trick might be to find drugs that target oncogene proteins. According to the oncogene theory, tumors depend specifically upon oncogene proteins for their cancerous properties. The growth of normal cells, which do not carry mutant oncogenes, should not be much affected by drugs targeting oncogene products.

I described my idea to my boss. He liked it and asked me to write it up as a formal research proposal. The company I was working for had no anticancer group, so they needed outside expertise to vet my idea. The corporate head of research was friends with Professor Sir Henry Harris at Oxford University and selected him to evaluate my proposal.

Sir Henry was the Regius Professor of Medicine at Oxford University and a cancer expert. The Regius Professorship of Medicine at Oxford is a highly prestigious appointment. In medieval times, in addition to being a professor at Oxford, the appointee was the king's personal physician. Sir Henry read my proposal and returned his assessment a few weeks after receiving it. The review was scathingly harsh. Sir Henry said that my idea was totally unworkable. He claimed that several of the scientific papers I cited in support of my idea had major flaws and were about to be retracted, and he described me as a young man attempting to work in areas far beyond the limits of my ability.

Despite the negative review, I wasn't terribly disappointed. Only about 1 percent of drug discovery projects make it all the way from conception to regulatory approval. The rest are derailed by technical, scientific, commercial, or financial issues and often some combination of these. I did not like Sir Henry's crack about me "trying to work in areas that were far beyond the limit of [my] ability," but I wasn't by any means a cancer expert and presumed Sir Henry knew things I was unaware of. I moved on.

Sir Henry died in 2014 at the age of eighty-nine. His biographers describe him as holding views on the oncogene hypothesis that were far outside the scientific mainstream. In one of his last publications, and counter to the then-prevailing scientific consensus on the matter, Sir Henry argued that "cancer is not caused by the direct action of oncogenes." The authors of two of the scientific papers I had cited in support of my idea went on to win the Nobel Prize in Physiology or Medicine. Today there are more than a hundred FDA-approved anticancer medicines that work via an action on an oncogene protein or process, and many more such drugs are currently under development. The oncogene hypothesis became accepted as Sir Henry and other like-minded scientists died off, a clear embodiment of Planck's principle.

Recognition most commonly occurs late in a scientist's career and only after a long struggle to convince one's peers that one's ideas are correct. This book describes the work of sixteen scientific innovators, women and men, many known to the general public and some perhaps not, who suffered the effects of Planck's principle. Science courses teach the facts, but generally little time is spent on how the facts were discovered and even less on the bizarre twists and turns that can lead from scientific inquiry to scientific knowledge, not to mention the great personal costs the process entails for individual scientists. All of the scientists included in this book went through years of struggle to get their ideas recognized and have their work become accepted, and then only extremely late in their lives. In some cases, acceptance came only after their deaths.

INNOVATORS

1 | Max Planck
Physicist who explained how light carries energy

We are taught in school that science progresses iteratively, with each new idea slowly building on the prior ones. In 1675 the physicist Isaac Newton, in describing his scientific achievements, wrote: "If I have seen further, it is by standing on the shoulders of Giants." But in the mid-twentieth century, historians of science came up with a new idea: that science instead progresses as a series of revolutions. A leader in this new thinking was Thomas Kuhn, a professor of the history and philosophy of science at Harvard, Berkeley, and Princeton. In 1962 Kuhn published *The Structure of Scientific Revolutions*, in which he argued that science progresses via abrupt shifts in how it views the world. He called these abrupt shifts scientific revolutions, changes that are "a noncumulative developmental episode in which an older paradigm is replaced in whole or in part by an incompatible new one." His theory is now generally accepted by the academic community. Kuhn explained that scientific disciplines are all guided by an overarching theory or model of how things work. He called this model the discipline's paradigm. Most of the time, scientific fields follow a process he called normal science. Normal science utilizes the current paradigm to design and execute experiments in order to solve puzzles in the field, puzzles that could not have been solved without the guidance of the paradigm. The results of these experiments confirm, support, and extend the paradigm. But this does not go on forever.

At some point, new results are obtained that conflict with the guiding paradigm. With time, many conflicting results accumulate, and these create a crisis in the field. Kuhn called this awkward period one of "revolutionary science." He explained that "scientific revolutions are inaugurated by a growing sense . . . that an existing paradigm has ceased to function adequately in the exploration of an aspect of nature to which that paradigm itself had previously led the way." For progress to occur, science employs the inadequacy of the old paradigm to provide the basis for the creation of the new one. In Kuhn's words, "Truth emerges more readily from error than from confusion." Eventually, a new paradigm is developed that can explain all findings. There follows a period of tense rivalry between the old and new paradigms, but over time the new paradigm finally takes hold and the discipline returns to normal science, solving fresh puzzles with the new paradigm.

A change in paradigm is an extremely rare event. First of all, it takes many years before conflicting scientific results accumulate to the point where the old paradigm is no longer tenable. And it then takes years until some genius comes up with the revolutionary new idea that explains everything. On top of that, a huge effort is then required to get scientists who have worked their entire lives conducting normal science under the old paradigm to accept the change.

Kuhn was a professor at Princeton during the time I was a biology graduate student there. Although students in the natural sciences, such as biology, commonly do not pay much attention to what is going on in the history and philosophy of science, Kuhn's book was popular among my classmates. The ideas in the book seemed compelling. After all, we were young trainees who had not invested much time working under the existing paradigms of our fields, so a change in paradigm was not of much concern to us. Some of my classmates likely fantasized that maybe someday they would be the one to lead a revolutionary change in their field.

But at the time, I think none of us appreciated the colossal effort required to carry out a scientific revolution, the years and years of frustrating, unrecognized work needed to convince the scientific community to make the change. Take a look at Nobel Prize awardees for physics, chemistry, or physiology or medicine. They are all old men and women. Nobel laureates in science are on average forty-five years old when they perform their Nobel Prize–winning work and wait an average of twenty-two years to actually get their prize.

In 1900 the physicist Max Planck came up with a controversial new idea that laid the foundation for the modern field of quantum mechanics. His idea was so new and unusual that it took almost two decades for other physicists to accept it as being true, and Planck waited eighteen years before the Nobel Prize Committee recognized the validity and importance of what he had done. In later life, he expressed his bitterness about the frustrating process that delays acceptance of new ideas in science, writing: "A new scientific truth does not triumph by convincing its opponents and making them see the light, but rather because its opponents eventually die and a new generation grows up that is familiar with it." This idea, often expressed succinctly as "Science advances one funeral at a time," has become known as Planck's principle.

Early Life and Studies

Max Planck was born in 1858 in Kiel, Germany, into a scholarly family. Planck's baptized name was Karl Ernst Ludwig Marx Planck, but he soon took Max, a variant of his middle name Marx, as his first name and used it for the rest of his life. His paternal great-grandfather and grandfather were both theology professors at the University of Göttingen. His father was a law professor at the University of Kiel and later a professor at the University of Munich. One of his uncles was a judge.

But no one in this family of academics was a scientist. Planck was encouraged to pursue a career as a scientist by Hermann Müller, his teacher of astronomy, mechanics, and mathematics at the Maximilians gymnasium, his high school in Munich. Planck studied physics at the University of Munich and later at Friedrich Wilhelms University in Berlin. In October 1878 he passed his qualifying examinations and then defended his doctoral dissertation, "On the Second Law of Thermodynamics" (*Über den zweiten Hauptsatz der mechanischen Wärmetheorie*), in February 1879. He went on to complete a habilitation (a German degree that is a type of second doctorate) based on the thesis titled "Equilibrium States of Isotropic Bodies at Different Temperatures" (*Gleichgewichtszustände isotroper Körper in verschiedenen Temperaturen*).

Planck started his academic career as a Privatdozent (German academic rank comparable to lecturer) in Munich and in April 1885 became an associate professor of theoretical physics at the University of Kiel.

Modern Physics

Planck worked as a physicist at a time when physics was beginning the transition from what we now call classical physics to modern physics. Classical physics describes and explains the composition and behavior of the world around us. It encompasses three ideas: 1) Newton's laws of motion and universal gravitation, which explain gravity, and his laws of conservation of energy and momentum, explaining the motion of objects and the forces acting on them; 2) the laws of thermodynamics, which explain how heat, work, and temperature are connected and how they relate to other forms of energy; and 3) Maxwell's equations, which explain electromagnetism, classical optics, and electric circuits.

Because classical physics describes and explains the composition and behavior of the world around us as we perceive it, it can be understood and visualized based upon everyday experience. Newton's laws of gravity and motion: you drop something off a roof, and it heads for the ground, accelerating as it goes. Easy to visualize and understand. Thermodynamics, the relationship between heat and other forms of energy: you boil water in a pot and soon the lid starts clattering. There is an obvious relationship between heat energy and mechanical energy: the heat energy turns into mechanical energy and lifts up the pot lid. Makes complete sense. Maxwell's equations on electricity and magnetism: you wrap a nail with wire, send a current through the wire, and the nail becomes magnetic. You probably did this in fifth grade. Electricity and magnetism are related. Easy to see, easy to understand.

But starting at about the turn of the twentieth century, physics transitioned into modern physics. Physicists began to ask questions about the composition and behavior of things we cannot see or experience, things that go unimaginably fast and are inconceivably massive, and things that are impossibly tiny, immeasurably smaller than anything we can imagine. It would have been nice if tiny objects and enormous entities, things that are moving at breakneck speed, all obeyed the same laws as the things with which we are familiar. But unfortunately, it turned out that the universe does not work that way. And that created problems for the physicists working on modern physics.

The Extremely Massive and Unimaginably Fast

The explanation for how things that are going unimaginably fast or are inconceivably massive behave was discovered by Albert Einstein and is called relativity. The science describing the physical laws governing how things that are impossibly tiny behave was originated by Max Planck and is called quantum mechanics.

Einstein's theory of relativity describes things that are true but that at the same time one's common life experiences say cannot possibly be true. *Time goes slower as you go faster.* How can that be? Time is time. It goes at a constant rate. *Things become more massive as you go faster.* How can mass change? Mass is mass. *Mass bends the space-time continuum and prevents light waves from traveling in a straight line.* Einstein had to invent the notion of the space-time continuum in order to explain how things work at extreme speeds and how things work for extraordinarily massive objects. His explanation meant that light does not always travel in a straight line, even though one's experience says that the path light takes is probably the best imaginable example of a straight line.

Prior to Einstein, there were conflicting physical observations in physics that no one could reconcile. For example, nothing can go faster than the speed of light and yet gravity is instantaneous. Both were known to be true based on well-accepted physical observations. But it seemed that they cannot both be true. The field was unnerved and floundering. The goal of physics was to explain the physical world, and physicists were stymied.

Einstein resolved the issue with his theory of relativity. As hard to imagine as it was, the theory settled the conflict. As no one could come up with any other idea to explain all observations, physicists soon came to embrace relativity, although sometimes kicking and screaming. Einstein published his theory in 1905 and received the Nobel Prize in 1921, reasonably quick acceptance for such a totally revolutionary idea.

But it still seems crazy. The average person may say, "Why should I care? Why am I even wasting my time reading this?" Crazy as it may seem, this is the way the world works. Take GPS, for an example. The global positioning system could not work if the engineers who built it did not take the theory of relativity into account.

For GPS to work accurately, the clocks on the GPS satellites and the clocks in GPS tracking equipment on Earth must agree with an accuracy of twenty to thirty nanoseconds. But the clocks on the satellites are moving at about 14,500 kilometers an hour (9,000 miles an hour). Because of the high speed of the satellite relative to the Earth, Einstein's theory of time dilation says that the clock on the satellite should tick more slowly, by about seven microseconds per day. Also, the satellite is in orbit high above the Earth's surface, where gravity is weaker. According to Einstein, in the weaker gravity the clock speeds up by about forty-five microseconds per day. Taken together, the satellite clock runs thirty-eight microseconds per day faster than the Earth clock in your cell phone GPS application.

Without correction, the entire GPS system would be off by ten kilometers after a day, seventy kilometers after a week, and 3,640 kilometers after a year. After a year without correction, the GPS system would tell you that your car, sitting comfortably in New York City, was actually in Salt Lake City, Utah. Not very useful. To account for this, the engineers who made the GPS satellites installed clocks that run slowly by just the right amount to counter Einstein's predicted relativity effects. And as a result, the GPS says your car is exactly where it is.

The Inconceivably Tiny

For the most part, Planck and Einstein worked at the opposite ends of physics. Einstein explained how things that are going unimaginably fast or are inconceivably massive behave. Planck showed how things that are impossibly tiny work. (Best known for the theory of relativity, Einstein also showed that the photoelectric effect is produced because light is at once a wave and consists of tiny packets of energy now known as photons. The brilliant Einstein thus studied the physics of both the unimaginably fast/inconceivably massive and the impossibly tiny.) But Planck did not intentionally set out to pursue the scientific findings for which he is now famous.

When Planck started his research career, his principal interest was in thermodynamics. While the laws of thermodynamics known in Planck's time gave a good general description of how heat, work, and temperature were related, there were still a few phenomena that could not be completely explained by or understood by thermodynamics. One of these was black body radiation.

Thermal radiation is spontaneously emitted by ordinary objects. When this emission is measured under a set of idealized and controlled conditions, it is called black body radiation. At room temperature, thermal energy radiates in the infrared spectrum. Since we cannot see infrared light, it appears to us like nothing is happening. But when an object becomes hotter, it first appears dull red because it is emitting radiation within the visual spectrum. We call this *red hot*. And as temperature increases further, it appears bright red, orange, yellow, white, and finally blue-white. We call this high-temperature radiation *white hot*.

There was a clear relationship between radiation frequency (color) and temperature, but how exactly did this relate to the amount of energy involved, and what was the explanation for the effect? The first scientist to tackle this problem was Wilhelm Wien in 1896, and his explanation is called Wien's law. But Wien's law correctly predicted the behavior of only high-frequency emissions. It failed at low frequencies. In 1905 the British scientists Lord Rayleigh and Sir James Jeans formulated the Rayleigh–Jeans law. But although the Rayleigh–Jeans law correctly predicted experimental results at low-frequency emissions, it created what became known as the "ultraviolet catastrophe" at high frequencies. It predicted that at high frequencies matter could instantaneously radiate all of its energy until it is near absolute zero. This does not happen, so the Rayleigh–Jeans law is wrong too.

Then Planck proposed what turns out to be the correct explanation. It is called Planck's radiation law:

$$E = h\nu$$

where E stands for energy, h is Planck's constant ($6.62607015 \times 10^{-34}$), and ν is the frequency of the emitted light. Superficially, this seems pretty simple and straightforward. The energy produced in black body radiation is a precise fraction of the light frequency. So far so good. But there is an uncomfortable and disconcerting implication in Planck's radiation law.

Scientists had long wondered what light was composed of. In 1801 Thomas Young carried out what is called the double-slit experiment. Young used a light source to illuminate a plate into which two parallel slits had been cut, and then observed the light on a screen behind the plate. The light passing through the two slits produced interference: the two light beams, seen as a series of bright and dark bands on the screen, alternately reinforced or canceled each other. In the experiment, the slits split a single light beam into two beams of light and introduce a time, or phase, difference between them. Depending upon their phases, two waves can either reinforce one another (most strongly when their phases are identical) or cancel each other (most strongly when their respective phases are 180 degrees out of phase). Particles have no phases and do not produce this effect. This result showed that light must be a wave, not a stream of particles, and for the next one hundred years physicists believed that light, X-rays, and radio signals were all waves.

But Planck's radiation law implies that light energy is emitted in tiny discrete energy packets of a specific size. Although Planck's equation, $E = h\nu$, says nothing directly about energy packets, in order for this equation to be true, light must be a stream of energy packets. Light could not exist as a simple, continuous flow of energy. These energy packets are now called photons. So, was light composed of waves or was it composed of energy packets? Planck's radiation law argued that it was both, but this is impossible to visualize. Planck's idea confounded common sense, yet it was true. The world of the impossibly tiny does not operate by the same laws that work in the world we inhabit. With his radiation law, Planck had founded quantum theory and entered the physics of the impossibly tiny through the back door.

The Greek philosophers were the first to speculate that matter is composed of tiny discrete units. Their writings were rediscovered by philosophers in fourteenth-century Europe, who continued to discuss and debate this concept. Later on, this idea was championed by chemist Robert Boyle and physicist Isaac Newton, but without experimental support. The idea that matter was composed of tiny discrete units received its first experimental backing at the end of the eighteenth century.

John Dalton was an English scientist who discovered the law of multiple proportions. He studied chemical oxides, compounds made from oxygen and another element, in his case tin, iron, and nitrogen. He found that there were multiple ways in which each of these elements could combine with oxygen: tin and iron each combined with oxygen in two ways and with nitrogen in three ways. But in every case, the combinations appeared in whole number multiples. For tin, oxygen combines in either one amount or twice that amount. Oxygen combines with iron in two possible whole number ratios, the larger of which we call rust. And with nitrogen, oxygen combines in either an amount, twice that amount, or four times that amount.

Dalton wrote that these whole number ratios were best explained if discrete, individual particles of oxygen were combining with particles of the other elements: one or two for tin, two or three for iron, and one, two, or four for nitrogen. Dalton called these particles atoms. He further proposed that each chemical element is composed of atoms of a single, unique type, that atoms cannot be altered or destroyed by chemical means, and that atoms can combine to form more complex structures. It was the first time atomic theory was proposed based upon experimental observations.

Atoms were initially thought to be fundamental units lacking substructure, and so the smallest possible division of matter. But in 1897 British physicist J. J. Thompson discovered a new particle, the electron. Electrons carried a negative charge and were 1,800 times smaller than the hydrogen atom, the smallest atom. Thompson proposed that atoms were composed of smaller structures and that the electron was one of these. Since electrons carry a negative charge but atoms are neutrally charged, there also had to be a component of the atom with a positive charge to neutralize the electron's negative charge. This is the positively charged nucleus.

Experiments over the next decade yielded the Bohr atom: a general model of atomic structure that is still generally accepted and taught in elementary and high schools. The Bohr atom was proposed in 1913 by the Danish physicist Niels Bohr and the British physicist Ernest Rutherford. Their model said that the atom has a small, dense, positively charged nucleus with tiny electrons flying around it. Our solar system provides a structure analogous to the Bohr atom, with the sun at the center and the planets revolving around it. This arrangement of the solar system provided a clear and solid analogy when the Bohr atom was proposed in the early twentieth century.

While the solar system is an excellent model for the physical structure of the atom, it is a terrible model for the atom's behavior. From the recent work of Planck, it had become clear that the laws governing the behavior of our own world were not necessarily going to apply to the atom. So when Bohr and Rutherford developed their model, they made sure that it was consistent with the quantized energy described by Planck's radiation law and incorporated the idea that electrons could not just randomly fly around the nucleus. Instead, they were fixed in specific orbital levels and at fixed energies. Electrons could on occasion change from one permitted orbit to another, and when they did so it would be an instantaneous "quantum leap" from one position to another, releasing or absorbing one of Planck's photon packets of energy.

This is, of course, not at all the way the solar system works. While planets are in orbits at fixed distances from the sun, there is no law saying that a planet in general has to revolve at only certain distances. In other solar systems, the planets are at different distances from their sun than the planets in our own solar system. And the laws of celestial mechanics do not permit planets to abruptly change their orbits after absorbing or releasing energy the way electrons do. Further work revealed even more extreme discrepancies.

With the acceptance of Planck's radiation law, the next generation of physicists worked to expand our understanding of the laws governing the world of the ultra-minuscule. Soon the field that had been established by Planck, the study of the world of the super tiny, became a subspecialty of physics in its own right, called *Quantenmechanik* in German and quantum mechanics in English. And as time went on, new quantum mechanics laws that were much harder to visualize and rationalize were developed.

The next generation of scientists made findings that defied everyday experience and common sense far more so than Planck's own work. In 1927 Werner Heisenberg published something he called the uncertainty principle. The uncertainty principle basically says that, in the world of the infinitesimal, you cannot simultaneously know both the position and the path of travel of a subatomic particle. You can know only one or the other. In fact, the very act of observing what is going on will alter the outcome. In our own world, imagine watching a baseball game, not knowing where the ball is or where it is going. Then, if we go about establishing where the ball is or where it's going, we thereby determine the outcome of the game.

Accepting that in the tiny world it was impossible to know where subatomic particles were and where they were going, physicist Max Born developed something he called the statistical wave function. The statistical wave function calculates the probability that a particle will be at a given point of space and time. Born made it possible to say how likely it would be for a subatomic particle to be in any particular place at any particular time despite not knowing these things for certain. Born was making educated guesses in a field that had always been based on precision and certainty.

And in 1925 another physicist, Wolfgang Pauli, showed what has become known as the Pauli exclusion principle. This law says that no two electrons in an atom can exist in the same quantum state. This was in real defiance of common sense. How could an electron possibly know what the other electrons were doing so that it could avoid being in the same quantum state as the other electrons? Were tiny electrons supposed to be communicating and coordinating their activities with one another? It would seem to be impossible, yet physicists even have a name for the effect: *quantum entanglement.*

These ideas were difficult for even professional physicists to accept. At a famous meeting in 1935, Albert Einstein debated the new quantum mechanics theories with physicist Erwin Schrödinger. Schrödinger, attempting to explain these new ideas, came up with the famous Schrödinger cat *Gedankenexperiment,* or thought experiment.

Schrödinger imagined a cat, a flask of poison gas, some radioactive material, and a Geiger counter in a sealed box. The radioactive material will randomly decay, and the resulting radio-activity is detected by the Geiger counter. The Geiger counter is attached to a hammer that will shatter the flask of poison gas when it detects radioactivity. This then releases the poison gas, killing the cat. A quantum mechanics analysis of this problem would say that the cat is simultaneously alive and dead. The paradox of whether the cat is actually alive or dead is resolved once one looks into the box. By so doing, one sees that the cat is either alive or dead, not both alive and dead. The very act of observation produces the outcome. This is how quantum mechanics works. While everything is statistical, at some point reality intervenes and produces one possibility or the other.

Again, one might ask, "Why should I care?" The result in Schrödinger's cat experiment is binary: either the cat is alive or dead; one or zero. It happens that all computing is binary, based on strings of ones and zeros. Therefore, in theory the principles of quantum mechanics could be applied to design a new type of computer. A number of groups are working on this around the world.

The attraction of quantum computing is that in conventional computers information is carried by electrons moving through circuits. The velocity of these electrons, as Albert Einstein was well aware, is linear and limited by the speed of light. But quantum effects are probabilistic. The speed of light does not apply. It should be possible to produce quantum bits in a massively parallel superposition of an enormous exponential number of states. Then all computations can be made simultaneously in parallel rather than consecutively. Thus, in theory, a quantum computer could be a lot faster than a conventional computer. Quantum computing is in its infancy relative to conventional computing, so it will take time to reduce all this to practice. In 2022 three physicists were awarded the Nobel Prize for their work in quantum technology, with the Nobel Committee stating that "ineffable effects of quantum mechanics are starting to find applications" including applications for "quantum computers, quantum networks and secure quantum encrypted communication." We will see what happens. If successful, computing speed will be dramatically increased.

Reactions to the New Ideas in Quantum Mechanics and an Example of Planck's Principle

Physicists took different approaches to dealing with these new physical laws that defied everyday experience. One approach was to rationally accept the laws, despite their strangeness and the fact that they were impossible to visualize and counterintuitive. Accepting them was also to admit that the principles that describe the world in which we live may not have universal application. A second approach was to accept the laws as models that explain the data but also to state that these laws do not necessarily have a physical reality. Two examples follow.

Light is composed of waves. Waves need a medium in which to propagate. Ocean waves need water. Sound waves need air. Light waves can travel in a vacuum, so what medium does light use? One group of physicists hypothesized that something exists that they called the luminiferous ether. Luminiferous ether is invisible and infinite and does not interact with physical objects. It behaves like a fluid in order to fill up space but also is millions of times more rigid than steel in order to support the high frequencies of light waves. It is massless and without viscosity and completely transparent, nondispersive, incompressible, and continuous at even the smallest scale. This group of scientists was so desperate to make light waves conform to our everyday experiences that they invented an impossible substance in order to explain an inconceivable phenomenon.

The other group, led by Einstein, said there was no luminiferous ether. They argued that light waves were special and could travel without the need for a medium to support their propagation. First of all, Einstein's special theory of relativity said there was no single universal frame of reference, undermining one of the arguments that demanded the existence of luminiferous ether.

In our frame of reference, a medium is required for the propagation of waves. In another frame of reference, a medium might not be needed. Both things could be true and also totally mutually exclusive. You do not have to like it to believe in it. Planck's work was particularly supportive of the non-ether point of view, for it said that light was a stream of particles that had a "wave-like nature." Particles obviously do not need a medium in which to travel (spacecraft move just fine in the vacuum of space), so thus neither does light. Luminiferous ether is not needed.

At the time disputes of this type were erupting in the physics community over all sorts of ideas. Ernst Mach was an Austrian physicist and philosopher who is best known today for his study of shock waves. The unit of velocity for supersonic aircraft, measured as the ratio of the speed at which the aircraft is traveling to the speed of sound in air, is called the Mach number in his honor. Mach and Planck disagreed about the reality of atoms. Mach said that atoms were mere theoretical constructs. The atom was just an idea that allowed physicists to make sense of their data. Planck, on the other hand, said that, as difficult as it was to visualize and conceive of solid materials being composed of trillions of tiny particles, atoms did in reality exist.

Some scientists remained neutral. These scientists said that a physical law that defied everyday experience could not be true. But a true explanation had to exist. Therefore, one had to wait until eventually some smart scientist came up with a hypothesis that explained the data and was at the same time consistent with how we see the world is working.

The new findings of the younger quantum mechanics physicists were too much for both Einstein and Planck. They could not accept them. In regard to Born's statistical wave function, Einstein famously argued, "God does not play dice." Planck remained adamantly opposed to the indeterministic, statistical worldview developed by the later quantum mechanics physicists. He rejected their ideas, saying they were in conflict with his deepest intuitions and beliefs. The physical universe, Planck argued, is an objective entity existing independently of man. The observer and the observed are not intimately coupled.

While both men had trouble accepting the new findings, their reactions were quite different. Einstein argued against the new ideas, but having voiced his objections, he moved on, working on his unified field theory and trusting that the scientific process would eventually sort out the correct answer. Einstein seemed to take delays in acceptance of new ideas in physics as the way of the world, or at the very least the way of science. Planck, for his part, remained adamantly and bitterly opposed to the new theories.

Objectively, it is hard to understand the basis for Planck's rigidity, dogmatism, and bitterness. Long delays in getting new ideas accepted, especially radical ideas, are common and standard. Planck first proposed his black body radiation law in October 1900 and published it in 1901. He was awarded the Nobel Prize for this work in 1918, a delay of seventeen years. Consider

Einstein in comparison. Einstein published his theory of relativity in 1905 and was awarded the Nobel Prize in 1921, a delay of sixteen years. Planck does not seem to have been singled out for poor treatment.

One would have expected that Planck, who was extremely resentful of his own treatment by the physicists who had rejected the quantum concept of his radiation theory, would have shown generosity and support for the new generation of physicists. But while taking his own treatment as a personal affront, he seemed incapable of empathizing with the plight of the younger scientists.

Perhaps tragedies in his personal life contributed to Planck's bitterness. Planck's first wife died in 1909 at the age of forty-eight, and his oldest son, Karl, died fighting during World War I. After the war, Planck's twin daughters, Margarete and Emma, both died in childbirth. And during World War II, his youngest son, Erwin, was implicated in the failed July Plot to kill Hitler and was hanged. In 1944, an Allied air raid hit his house and destroyed many of his possessions, including all his scientific notebooks.

After Hitler came to power, Planck decided to remain in Germany and tried to help Jewish scientists but had little success. To protest the religious intolerance of the Nazis and their persecution of the Jews, he resigned his presidency of the Kaiser Wilhelm Society and thus lost his job as the head of one of the most prestigious research institutions in the world. In the long run, his resignation protected him. During World War II, weapons and medical research was performed by the Kaiser Wilhelm Institute, including human experimentation on prisoners in Nazi concentration camps. The institute collaborated with Dr. Josef Mengele, who provided them with Jewish bodies and body parts for their research. As the American forces closed in, Planck's successor,

Albert Vögler, committed suicide, realizing that he had committed and would be prosecuted for war crimes.

Planck's last publication was his autobiography. In it he railed against those who delayed the acceptance of his ideas and proposed what is now called Planck's principle: that science progresses one funeral at a time; that a new idea in science cannot take hold until its opponents die off. His last contribution to scientific debate was to oppose intransigently the ideas of Born, Heisenberg, Pauli, and Schrödinger. All four went on to win Nobel Prizes as Planck had for his work, and their theories are now the standard principles of quantum mechanics, taught in physics classes at universities all over the world.

Ironically, Planck's career demonstrates Planck's principle in two totally different ways. In his early career Planck was frustrated by the unwillingness of older scientists to accept his work. And then, most unexpectedly, in his later career Planck became the rejecter, unwilling to accept the new ideas of the next generation of physicists. By becoming the rejecter, he provided a second validation for Planck's principle. Plank died refusing to accept modern quantum mechanics. As he had observed, scientific truths do not triumph by convincing the opponents but rather because the opponents die off.

2 | Gregor Mendel
Founded the science of genetics

We all know someone who was enormously talented but whose family lacked the resources to provide the training needed for that person to utilize their gift. Likewise, we all know someone who is highly capable but whose temperament prevents them from making it through an oral exam and convincing the examiners of his or her competence. And we all know someone who has something of value to offer but lacks the salesperson's skill to market it to others. Gregor Mendel was all of these people. Yet amazingly, despite these hurdles, he is the only person in the modern era to have established a completely new scientific discipline.

Physical traits of plants and animals are inherited, and animals and plants generally resemble their parents. The Greek philosophers Hippocrates and Aristotle wrote about heredity, but they did not explain how heredity works, nor did they predict what would happen as the result of any particular mating. Exceptions to the principle that offspring resemble their parents are commonplace, and any explanation of heredity would have to account for that. Two people with light eyes and light hair may have a child with dark eyes and hair. Two people with dark eyes and hair may have a child with light eyes and hair. For most of human history, the process appeared so random and haphazard to those who considered it that it could not be explained.

Gregor Mendel was born in 1822 in Hyncice, a small Silesian village that is now located in the Czech Republic. His parents were farmers, working land that had been in the Mendel family for more than 150 years. Gregor loved science and did not want to be a farmer, so he joined the Augustinian Friars, who were known for providing educational opportunities for their monks. After graduating from secondary school, Mendel first studied physics at the University of Olomouc and then, in 1851, transferred to the University of Vienna to study physics under Christian Doppler.

Doppler was one of the outstanding physicists of the nineteenth century, renowned today for explaining an important physical phenomenon that was later named after him: the *Doppler shift*. You may not know what a Doppler shift is, but we are all familiar with this effect. When a car goes by you sounding its horn, the pitch of the horn changes as the car passes: *haaaank* as it approaches you and *hoooonk* as it passes you. The pitch of the horn goes from higher to lower. Through Doppler's work we know that the actual sound of the horn does not change. Instead, the frequency or pitch of the horn changes for the observer as the result of the motion of the source of the sound relative to the motion of the observer. For the driver of the passing car, the pitch of the horn is steady because his or her motion relative to the horn is constant.

Mendel was an excellent scientist and an excellent student, but he was also extremely nervous and a terrible test taker. Prior to attending the University of Vienna, Mendel had worked as a substitute high school teacher with the intention of being certified to teach full time, but he failed the oral part of a three-part certification exam. He knew his stuff but fell apart under the pressure of the oral exam. After two years of study under Doppler, Mendel returned to the Augustinian St. Thomas's Abbey in Brno in 1853 to repeat the examination. He had learned a tremendous amount of science from Doppler, yet he again bombed the oral part of the exam. A scientific career was not going to be in the cards for Mendel, so he became a common friar at the monastery.

Still, Mendel was dissatisfied and wanted to continue doing science. Realizing he would never be able to do physics research at the monastery, he came up with an alternative idea that he pitched to the abbot. Monks at the Augustinian St. Thomas's Abbey grew their own food, as was typical for all monasteries at the time. Mendel argued that, since the mission of the abbey was biblical study, meditation, and prayer, any time spent growing food was secondary and detracted from this prime mission. However, if he could use science to increase crop yields, the monks would be able to redirect manpower away from farming and back to the prime mission: biblical study, meditation, and prayer.

The abbot accepted the argument and gave Mendel a plot of land and a gardener to conduct experiments. Mendel selected the common garden pea, scientifically called *Pisum sativum*, as the basis for his work. It may seem strange that Mendel selected this unimportant, commonplace plant, but there was a sound strategy behind his thinking. The garden pea is a member of the bean family. Monks ate a largely vegetarian diet. Vegans will immediately know why this was important.

Grains, for example, those used to make bread, provide incomplete nutrition because they do not contain adequate quantities of the amino acid lysine. If people eat only grains, they become malnourished. You must supplement your grain diet with a food rich in lysine. Beans are the answer, but beans do not contain adequate quantities of the amino acid methionine, so you can't eat just beans, either. Beans have plenty of lysine, and grains have plenty of methionine; therefore, if you eat a diet composed of both beans and grains you will not be malnourished. Today we call this *protein complementing*.

All human societies have dishes that combine a bean and a grain. Native Americans ate succotash: corn and lima beans. In Latin America, people eat beans and rice; in the Middle East pita and falafel; in Asia rice and bean curd. Mendel was focusing on the bean part of the diet.

Mendel started out as any good scientist should. Just as they are today, seed varieties were a commercial product in Mendel's time. Seed dealers would compete with one another by offering different seed varieties with characteristics that would be desirable to growers. Mendel ordered thirty-two different pea seed varieties and planted them in his garden for several growing seasons to make sure that they bred true, that is, that the offspring of each variety were always identical to the parents. (Commercial seeds are highly inbred and give rise to uniform, identical offspring, but Mendel, like a good scientist, did not believe the dealers. He wanted to make sure for himself so he could be confident that any variation he later saw had to be the result of his own experimental manipulations.)

Despite the seed dealers' claims to the contrary, not all the varieties Mendel purchased bred true: some produced variants in subsequent growing seasons. Mendel then selected several of the varieties that had proved out to be true breeding and used them to make hybrids. That is to say, he would cross one variety with a different variety to see what would happen. Mendel was not a biologist. He had been trained as a physicist, so he approached the problem like a physicist, not like the biologists of his time. In each cross he focused on a single trait. When he crossed a tall plant with a short one, he worried only about the height of the offspring. When he crossed a plant with yellow peas with one with green peas, he worried only about the pea color of the offspring. Biologists of the time would not have done that.

Nineteenth-century biologists worked qualitatively. They described things. Their approach would have been to make a complete description of the resulting offspring plants, not a single characteristic. Physicists in contrast were quantitative. They counted things. In narrowly focusing on specific traits, Mendel selected properties he could count. How many of the resulting plants were tall and how many were short? How many peas were yellow and how many green?

Mendel also thought like a physicist in coming up with a hypothetical model to guide his thinking. He postulated that traits were produced by hereditary factors. The height of a plant was produced by one as yet unknown hereditary factor. Pea color was produced by another as yet unknown hereditary factor. He wanted to determine how his hypothetical hereditary factors behaved. Today we call these factors *genes*, but the word *gene* wasn't invented until 1909, twenty-five years after Mendel's death.

Mendel conducted his experiments, made his crosses, collected and planted the hybrid seed, and tended the plants grown from the seeds until the end of the growing season, when he got his results. The result of his first experiment was not at all what one would expect. His tall-short pea cross yielded only tall plants. His yellow pea versus green pea cross yielded only yellow peas. What happened? The growing season was over and Mendel had until spring to think about his results.

Over the winter he came up with the idea of *genetic dominance*. He argued that each parent plant had successfully donated its own genetic factors to the offspring plants. But because tall was dominant to short, he saw only tall plants, and because yellow was dominant to green, he only saw yellow peas among the offspring.

But if his factor idea was correct, what happened to the short factor in the tall-short cross and what happened to the green pea factor in his yellow pea–green pea cross? Were the short and green pea factors lost? Had the tall factor destroyed the short factor and the yellow factor destroyed the green factor? Or were these factors somehow only suppressed but still existed, perhaps to be seen later? Mendel designed his next experiment to answer that question.

The next spring he crossed the hybrid progeny of the tall-short cross in brother-sister matings, and the hybrid progeny of the yellow pea–green pea cross in similar brother-sister matings. (Don't worry, plants don't care about incest.) The following spring he planted the seeds that had resulted, and by that fall he had his answer. Short plants reappeared, as did green peas. The hereditary factors for these traits had been passed on. So logically, for this to have happened, there had to be multiple copies of each hereditary factor in the plants: at least one short and one tall and at least one yellow and one green. Mendel's later mathematical analysis showed that each plant must carry two hereditary factors (one from each parent).

What was the story with the hereditary factors present in the hybrid plant whose traits were unseen? Mendel called such hereditary factors recessive, the mirror image of the dominant. When the dominant hereditary factor was present, the recessive factor did not contribute to the plant's appearance. The recessive factors were still there, but their appearance in the hybrid plants was somehow suppressed.

And then Mendel did something that no other biologist had ever done before. He counted all his resulting plants. Here again, he was thinking like a physicist, not a nineteenth-century biologist. When he was done, he had a strange and difficult challenge to explain the result. The offspring appeared in three-to-one ratios. There was one short plant for every three tall plants and one plant with green peas for every three plants with yellow peas. Why?

Again, the growing season was over, and Mendel had a long winter to ponder his results. That winter he was supposed to make up for time lost to experiments by spending extra time meditating, but historians think he was only pretending to meditate. Instead, he was asking himself, "Why a three-to-one ratio?"

By the spring he had worked out a mathematical solution. Mendel reasoned that each parent plant had two factors for each trait and would give one of these at random to each offspring. In pure-breeding plants, these factors were identical and thus all the offspring received identical factors. But hybrids, like those with tall crossed with short parents, had two different factors. Because of the law of dominance, they all appeared to be tall. And now came Mendel's great insight.

When hybrids were mated to one another they produced four possibilities. If they inherited two short factors, they appeared short. If they inherited two tall factors, they were tall, just like the true-breeding original tall variety. Lastly, there were two combinations to make plants that were just like their hybrid parents and these, just like the parent plants themselves, appeared tall. Thus, there were three ways to make a tall plant and only one way to make a short plant. Three to one, just like the data said.

Thus far, Mendel had been looking at one trait at a time. But he wondered what would happen if he looked at two traits. Would the hereditary factors be somehow tied together or would they each go their separate, independent way? He set up an experiment to answer this question. He mated a true-breeding variety with yellow, round peas with a variety that had green, wrinkled peas. He already knew that yellow was dominant to green and that round was dominant to wrinkled. So, in complete accordance with the genetic dominance, the resulting hybrid plants made only yellow, round peas. And then the following growing season he did brother-sister matings. The results came out the following fall.

Every combination of yellow, green, round, and wrinkled was present. By counting all the types of offspring it became clear that these two sets of hereditary factors were not tied together. They behaved in a totally independent way as they passed from generation to generation. We now call this *independent assortment*.

Current biology students sometimes question this last result. Today we know that genes are on chromosomes. So, genes that are on the same chromosome should be linked together and move together in crosses. But Mendel saw no such linkage. Students commonly dismiss this concern by thinking that Mendel must have been fortunate to have chosen genes that were on different chromosomes. Since they were on different chromosomes, his genes would be expected to assort randomly. But if he had been unlucky in selecting genes that were on the same chromosome, he might not have discovered independent assortment.

Mendel studied a total of seven genes. From modern genetic studies, we know that the garden pea has seven chromosomes. One could imagine that Mendel by chance could have chosen genes on different chromosomes. But modern genetics has shown that among the seven genes Mendel picked, two are on chromosome number 1 and three are on chromosome number 4. So how did Mendel get his result? Did the great scientist make a mistake?

Modern genetics has also shown that homologous chromosomes exchange pieces during the production of sperm and eggs. This exchange happens frequently. As a result, only genes that are adjacent or located very close to one another are inherited together. Genes that are on the same chromosome but located far apart get mixed up and assort independently, exactly what Mendel had found.

More than anything, Mendel's work demonstrated the power of a great idea. Mendel had no special training in biology or mathematics. He had no special knowledge of plants or agriculture. Knowledge of cell division or chromosomes, let alone DNA, lay in the future. Really, anyone could have done what Mendel did. No special equipment or procedures were needed. It was mainly just breeding plants, looking at them and counting. Hippocrates and Aristotle could have done it, but they did not think of it, so they did not try. My great-grandparents, who lived in Europe at the time, could have done it, but they did not think of it and did not try.

The genetic experiments Mendel did with pea plants took him eight years to complete. He then wrote up his results and interpretation and presented his paper, "Experiments on Plant Hybridization" (*Versuche über Pflanzenhybriden*), at two meetings of the Natural History Society of Brno in Moravia in February and March 1865. Next, he published his results in 1866 in the scientific journal *Proceedings of the Natural History Society of Brünn* (*Verhandlungen des naturforschenden Vereines*). As scientific journals go, this was pretty insignificant, published by an obscure scientific society far from cosmopolitan centers of higher learning. But Mendel was just a common friar and was therefore probably lucky to get his work published at all.

Mendel knew what he had done was important, though, and he felt that good, competent scientists would quickly recognize the significance of his discovery once they read his work. He assumed that at least some good scientists might have encountered his published paper, but then nothing happened. There was no recognition. As best can be determined, almost no one contacted him. No one wanted to meet with him to discuss his ideas or congratulate him. No one started new projects to follow up on Mendel's work. Imagine if Neil Armstrong had stepped onto the surface of the moon and no one noticed. Why didn't anyone recognize the importance of what Mendel had done? Mendel thought perhaps he had missed something, some problem that others with more experience in biology had seen.

From the publisher's business records we know that the *Proceedings of the Natural History Society of Brünn* had 115 subscribers, mostly in Europe and chiefly universities or scientific societies. The publisher's records also show that Mendel received "40 *Separatabdruck*" (reprints). In the days before electronic PDF copies, scientists were given reprints, extra printed copies of their papers, to share with their colleagues. These reprints were especially important for low publication volume journals. The reprints would be sent by the author to scientists working in the field to ensure that colleagues were made aware of the new work.

It would have been standard practice for Mendel to send these reprints out. But historians have been able to trace only four of the forty reprints Mendel received: one was sent to Carl

Nägeli, one to Kerner von Marilaun at Innsbruck, one to a recipient whose name has been lost but who then sent it to Martius Wilhelm Beijerinck, who in turn sent it to Hugo de Vries, and the last was sent to an unknown recipient who apparently then gave it to Theodor Boveri, who then donated it to the Kaiser Wilhelm Institut für Biologie in Berlin, which in turn donated it to the Max Planck Institute in Tübingen.

Mendel had written to some of the great scientists of the day, enclosing a reprint of his paper, and asking them what they thought. As they were big shots, their correspondence files were preserved for posterity. The only celebrity scientist known to have replied to Mendel was Nägeli, and his response was negative and critical.

In 1866 Carl Nägeli held the prestigious post of senior professor of botany at Ludwig Maximilian University of Munich. He had rock star status within his profession and was the type of scientist who would be contacted for comment when an important newspaper like the *Frankfurter Allgemeine Zeitung* ran a news story about some new scientific development.

Nägeli's response was polite but discouraging. A major problem Nägeli saw in Mendel's research was that he was working with peas. In essence, he told Mendel that peas were not where the action was in biology and suggested that he switch to *Hieracium*, also called hawkweed, a type of sunflower. That was the plant Nägeli himself was working on.

Mendel was not a biologist and had no idea which species biologists thought were the best experimental systems. Plus, he had the problem of convincing the abbot to allow him do his research and for that reason had proposed peas, as an important source of nutrition for the monks, regardless of whether it was a highly valued experimental system. The monks were not going to eat a weed, and certainly not hawkweed. In any event, we now know that the rules of genetics are universal, so almost any species would have given Mendel the same result.

Most importantly, it is clear that Nägeli did not at all understand what Mendel was attempting to do. Mendel was not just trying to make some original observations about peas, which Nägeli obviously presumed to have been Mendel's favorite plant species. Instead, he sought to develop new, universal principles of heredity, and for that it did not matter which species of plant one chose to work with. The scope of this obscure monk scientist's ambition was totally lost on Nägeli.

Today, scientists who study Nägeli's life work believe he accomplished almost nothing of long-standing importance. Nägeli is mainly remembered by the scientific community for not understanding Mendel's principles of heredity and for being dismissive in his dealings with him. Mendel, for his part, was completely discouraged by the great scientist's response as well as the silence of others. He had toiled for more than eight years, and it looked like no one in the scientific community thought what he had done was worthwhile.

In the spring of 1868 Mendel was appointed as the new abbot of the Augustinian St. Thomas's Abbey. He took over the administrative duties of the abbey and never again did any scientific work. Church records document that Mendel was an able and competent administrator and was well liked by the friars who worked under him. When Mendel died in 1884 at the age of sixty-one, he was given a moving eulogy by the church fathers, and the well-known Czech composer Leoš Janáček played the organ at his funeral.

That would have been it, but as time went on biology began to change. More and more biologists began to think quantitatively, like physicists. In 1900, thirty-four years after Mendel's paper had been published and sixteen years after his death, three new scientific papers appeared, written by three scientists who had been independently studying heredity employing the new quantitative approaches: Carl Correns, Erich von Tschermak, and Hugo de Vries.

Carl Correns was a German botanist and geneticist and a professor at the University of Tübingen. Ironically, Correns had been a student of Carl Nägeli, the scientist who is famous for having been highly critical of Mendel's work. But biology was changing rapidly, and Correns came to reject some of Nägeli's teachings, took a quantitative approach to his own work, and even used the garden pea, a species that Nägeli regarded as a poor choice for biological research as an experimental organism.

Erich von Tschermak was an Austrian agronomist and a professor at the University of Agricultural Sciences in Vienna. His research focused on the development of disease-resistant wheat, rye, and oat crops. In developing disease-resistant varieties, von Tschermak carried out experiments to examine the heredity of traits in these plants.

Hugo de Vries was a Dutch botanist and professor at the University of Amsterdam, where he conducted a series of hybridization experiments, employing varieties of multiple plant species.

Correns, von Tschermak, and de Vries all reported essentially the same basic scientific observations and conclusions that had been described in Mendel's 1865 paper. The first of this trio of papers to appear was de Vries's. Martius Wilhelm Beijerinck, a professor at the University of Delft and de Vries's colleague and friend, read an early copy of his paper and wrote to him saying that he had come into the possession of a reprint of a paper by an obscure scientist named Mendel. It sounded like something de Vries would be interested in, and Beijerinck sent this reprint to him so he could check it out.

How de Vries reacted when he read Mendel's thirty-four-year-old study is not recorded. I think most people would have been in shock. Having worked on his project for more than ten years, de Vries clearly understood that he had discovered something of huge importance to science. And he probably believed he had likely founded a new field of science. But as it turned out, he had not. He had only confirmed observations that had been made more than three decades earlier by an obscure Augustinian friar.

To de Vries's enormous credit, he acknowledged Mendel's priority. A dishonest person might have written back to Beijerinck saying that a close reading of Mendel's paper showed that it was unrelated to de Vries's own work. Scientists do not always carefully read papers outside of their own field, and Mendel's paper was so revolutionary that it was difficult to read. (It is difficult to read even today because Mendel had to invent his own terminology, terminology that is not currently used.) But as a result of de Vries's honesty, today the fundamental principles of genetic inheritance are called Mendelian genetics and not de Vriesian genetics. Mendel, of course, never knew. He died in relative obscurity. And without the careful and honest reading of the scientific literature by a few scientists a decade and a half after his death, no one would have ever known. Mendel was so far ahead of his time that it took three and a half decades for scientists to be able to appreciate what he had done, and he died thinking he was a scientific failure.

The papers by de Vreis, von Tschermak, and Correns appeared at a time when the field had changed dramatically. Biologists of the early twentieth century were ready, and eager, to comprehend and appreciate what they had written. Mendel had learned four things: 1) hereditary factors come in pairs, 2) they are inherited at random, 3) some hereditary factors are dominant and some are recessive, and 4) the hereditary factors move from generation to generation separately and independently from one another. That was a lot.

All of modern molecular biology and much of modern biomedical science was developed based upon the discoveries Mendel made. Today people casually talk about recessive genes as if this concept were the most obvious thing in the world. Without Mendel's discovery of the gene, there would be no monoclonal antibody drugs (drugs like Humira to treat rheumatoid arthritis, Tysabri treatment for multiple sclerosis, and Herceptin treatment for breast cancer), no recombinant DNA drugs (drugs like human insulin for diabetes, growth hormone for dwarfism, and moroctocog alfa and similar drugs to treat hemophilia), no COVID-19 PCR diagnostics, no understanding of inherited diseases. Mendel was not a failure. He was a father of modern biomedical science.

The scientific community did not accept the new ideas put forward by Planck and Mendel, but for totally different reasons. Planck's ideas were rejected by the physics community, despite the fact that they were correct, because they defied everyday experience and common sense. Mendel's ideas, equally correct, differ in that they are totally reasonable and appeal to common sense. Mendel did not challenge the scientific status quo. He just discovered something new. So why were his ideas rejected by the biology community? Like all human endeavors, science runs on sociology as well as rationality. Mendel was just a common friar with no scientific credentials. The scientific establishment could not accept the possibility that a rank amateur could discover something that had eluded the most highly trained, most brilliant minds of the era. They therefore ignored him.

3 | Barbara McClintock
Discoverer of transposons (jumping genes)

Perhaps great geneticists are jinxed. Almost ninety years after Gregor Mendel began his experiments with peas, thus establishing the new field of genetics, another geneticist, Barbara McClintock, started her own genetic investigations with corn, only to suffer rejection and disbelief just as Mendel had almost a century earlier. Mendel's problem was that he was a true outsider, someone who, despite lacking training and credentials, went on to pursue a program of innovative botanical research. McClintock in contrast had received formal training in genetics at well-regarded research institutions. But she was a woman in a man's field whose colleagues denied her the opportunities, support, recognition, and respect she deserved.

Early Life, Education, and Research Appointments

Barbara McClintock was born in Hartford, Connecticut, in 1902, where her father practiced medicine. Despite the objections of her mother, who feared that too much education would make a woman unmarriageable, McClintock continued her education at Cornell University's College of Agriculture after she graduated from high school.

At Cornell McClintock developed an interest in genetics through the encouragement of her professor, C. B. Hutchison, a well-known and highly respected plant breeder and geneticist. McClintock stayed at Cornell to earn a BS, MS, and PhD in plant breeding and genetics. After earning her PhD, she was awarded several postdoctoral fellowships from the National Research Council, allowing her to continue her studies at Cornell and later at the University of Missouri and the California Institute of Technology. In 1933 a Guggenheim Foundation grant fund allowed her to pursue six months of research in Germany with the geneticist Richard B. Goldschmidt at the Kaiser Wilhelm Institute for Biology in Berlin.

Had she been a man, McClintock would have rapidly found a tenure-track professorship at a major research-based university after finishing her training. Smoking a pipe and wearing a Harris Tweed sport jacket, she would soon have been leading students in intellectual discussions in small group recitation sections. But in the early twentieth century, there was no established career path for women scientists, even for top-notch scientists with first-rate intellects like McClintock. So instead, after a brief return to Cornell, McClintock took an assistant professorship in the Department of Botany at the University of Missouri that had been specially created for her by geneticist Lewis Stadler.

McClintock soon became disappointed with her position at the University of Missouri. She was excluded from faculty meetings, was not informed of professional opportunities, and felt overly dependent upon Lewis Stadler, since she had not joined the university by the normal route and instead had been hired as a result of Stadler's influence and personal support. She also believed the department would never grant her tenure and lost

faith in the department senior faculty and the university administration. McClintock started looking for other opportunities and
in 1941 accepted an appointment as a staff scientist in the genetics department of the Carnegie Institution's Cold Spring Harbor
Laboratory, where she would remain for the rest of her career.

Where Are Genes Located?

When Gregor Mendel established the new field of genetics, he said
that the inheritance of traits was determined by hereditary factors
within the cells, factors that were later named *genes*. But Mendel
did not discover what the genes were made of, nor did he determine where they were located. Determining these things was left
to later scientists. In 1902 Theodor Boveri reported that normal
embryonic development in sea urchins required chromosomes to
be present, and Walter Sutton reported that during cell division
the chromosomes behaved in precisely the way that genes would
be expected to behave. Based on this, they proposed, but without
any evidence, that genes were likely located on chromosomes. By
simply counting the number of chromosomes and the number of
known genes, it was clear that there were many more genes than
chromosomes. Therefore each chromosome had to carry multiple
genes. But proving that this was true was another matter.

In the early twentieth century Thomas Hunt Morgan was
a biology professor at Columbia University. He had studied
embryology, but the rediscovery of Mendel's laws in the 1890s
had totally upended biology. Suddenly there were new biological principles, new strategies for experimental design, and new
possibilities to solve problems that had long stymied biological
scientists. Morgan decided to switch to genetics. At that time, all
genetics research was done with plants. But Columbia is located

in the Morningside Heights section of the very urban borough of Manhattan, New York City. There were no fields where Morgan could grow plants to perform genetic experiments.

It is said that one day Morgan was walking down Broadway on his way to teach a class when he passed by a fruit vendor's cart. The cart was buzzing with tiny insects: fruit flies, scientifically called *Drosophila melanogaster*. Morgan realized that fruit flies were small enough to use in conducting genetic experiments even in a tiny laboratory in densely populated Morningside Heights. Morgan had his new experimental system for studying genetics.

In his early research, Morgan quickly discovered a mutant gene that produced white eyes instead of the normal red eyes in fruit flies. Morgan performed genetic crosses to see how the white-eyed gene was inherited. He crossed a white-eyed male with a red-eyed female. All the offspring had red eyes, a typical result indicating that the red-eye gene was dominant. Then he crossed a white-eyed female with a red-eyed male. All the female offspring had red eyes, but all the male offspring had white eyes!

Morgan had discovered a new mutation that behaved differently from any previously described mutation. In the past all sorts of mutations had been found that modified physical traits. Morgan's new mutation modified a trait (eye color), but it diverged from all prior discoveries in that it was inherited differently in males and females.

In 1905 American geneticist Nettie Stevens (working with beetles) and American zoologist Edmund Beecher Wilson (working with squash bugs and broad-headed bugs) independently showed that female insects had two X chromosomes and male insects had an X and a Y chromosome. This is also true in fruit flies, and in the early 1920s Theophilus Painter showed that it is also true in humans.

After performing a number of follow-up experiments, Morgan determined the following. His experiments showed that the white-eyed gene was recessive and was located on the X chromosome. Females have two X chromosomes, so, for the fly to have white eyes, both X chromosomes have to carry the white-eyed gene. But males have only one X chromosome. They therefore needed only one white-eyed gene to have white eyes. Morgan proved that the white-eyed gene was on the X chromosome. Wherever the X chromosome went, the gene went too. No other explanation fit the data. And he had provided a proof that genes were located on the chromosomes. Morgan received the Nobel Prize for this work in 1933.

But the X chromosome might have been a special case. Maybe the X chromosome is unusual. What about all the other chromosomes, the autosomes, which constitute the bulk of the chromosomes in every species? Humans have twenty-three pairs of chromosomes, and only one of these pairs is the XX or XY pair. Did the nonsex chromosomes also carry genes? It would be nice to have a more general proof, showing that genes were on these other, much more numerous nonsex chromosomes.

Barbara McClintock specialized in cytogenetics, the microscopic study of the chromosomes. She used corn, *Zea mays*, as her experimental organism. In 1931, working at Cornell with a graduate student, Harriet Creighton, McClintock came across a chromosome with an unusual structural feature. This chromosome had a tiny knob on one end that could be clearly seen in the microscope. McClintock reasoned that genes located on that chromosome might be inherited along with the little knob. She and her student set up genetic crosses and showed that certain genes were indeed inherited along with knob. She concluded that these genes must be physically located on the chromosome with the little knob because the genes moved together with the knobbed chromosome. This was a clear physical demonstration that genes were located on autosomes, the nonsex chromosomes, and not just on the sex chromosomes.

In 1931 McClintock and Creighton published their findings showing that genes are located on chromosomes. While Morgan received the Nobel Prize for his sex chromosome experiments with fruit flies, McClintock's work received less recognition. Perhaps this was because Morgan had made the initial finding, even though McClintock's finding was more general, as the vast majority of genes are not located on sex chromosomes. Or perhaps it was because Morgan was a tenured professor, while at the time McClintock had only a temporary appointment as a botany instructor. And perhaps it was because McClintock was a woman working in a field dominated by men.

Transposons (Jumping Genes)

McClintock continued her genetic studies in corn after she moved to Cold Spring Harbor Laboratory in 1941. In 1944 her research turned to a new direction. She began studying the genetics of a variety of mosaic corn, a breed of corn composed of two genetically different types. This strain of corn had kernels with different color patterns, as though the ears of several different corn varieties had been chopped up and then pieces assembled at random to produce the ears of this corn. And even more difficult to explain, the color patterns would vary from generation to generation. Over the generations new kernel colors would appear, only to disappear in a later generation.

As an expert in cytogenetics, McClintock started by studying the chromosomes in this strain. She found that one of the chromosomes, chromosome 9, was often broken and always broke at the same place. She named the site of the break *Ds*, the dissociation locus. Further studies of the Ds genetic element showed that it could change position on the chromosome. Additional studies with Ds showed that chromosome breakage required a second genetic element, which she called Ac, or activator. She found that the Ac genetic element could also appear in different places in the genome.

By the 1940s one of the most basic, established principles of genetics was that genes always appear on chromosomes in a fixed order, like beads on a string. And it was also thought that genes had to be stable in order to produce the consistent inheritance patterns that had been studied for more than a century. Ds and Ac changing positions was heresy. No one had ever previously reported anything like this.

One possible explanation was that these effects were the result of mutations. By 1944 mutagenesis was a well-understood process. Hermann Muller was soon to be awarded the Nobel Prize for his work in describing mutagenesis, a prize he received in 1946. But McClintock's data could not be explained by mutagenesis. First of all, mutations are rare. The effects McClintock was seeing were far too frequent to be the result of mutagenesis. And while the mutation rate can be accelerated by treatment with a mutagenic agent like X-rays, McClintock's plants had not been exposed to a mutagen.

As uncommon as new mutations are, they are also stable. Mutations only very, very rarely revert back to the original form. But reversion happened frequently in McClintock's corn strain. Furthermore, mutation was known to be able to cause a group of genes to flip 180 degrees on a chromosome but never to make a single gene move. Mutagens like X-rays cause chromosome breakage. But X-ray chromosome breakage occurs at random places, never repeatedly in the same place. Mutation was clearly not responsible for what McClintock was seeing.

So what was causing the effects she had observed? McClintock came up with a new explanation. She said that the genes she was studying were moving around, jumping from place to place randomly. This idea, however radical, could explain everything. It would explain why she was seeing her Ac and Ds genetic elements in different places. And the chromosome breaks she was seeing would be explained if sometimes when Ds landed it caused the chromosome to break at the landing point.

It would also explain the mosaicism she had started to study in the first place. She hypothesized that Ac would randomly land next to or within another gene. When Ac did this, it would affect the function of the adjacent gene at the landing place, changing the kernel color. When Ac would jump out again, this would restore the original kernel color. And lastly, the jumping was frequent, far more frequent than a mutation. It was a single idea that explained everything. The problem was that it was a radical idea, an idea that went against a lot of, if not all, conventional thinking in the field.

McClintock published her findings in 1950 and followed up the publication with a talk on her work at the 1951 Cold Spring Harbor Symposium. After any scientific presentation, there are always questions from the audience. Scientists in the audience will also commonly make helpful suggestions to try to improve and expand on the presented research project. After the presentation of some radical new idea, there is always vigorous debate, with scientists in the audience questioning the observations and interpretations that have been presented and offering alternative, often more conventional explanations. But one of the attendees at McClintock's talk later recalled that when she had finished speaking there was dead silence. No comments or questions from the audience. The audience was stunned by what McClintock had just said that went totally against all conventional thinking in the field. Heredity depends on an organism's traits, determined by genes, to be accurately handed down from generation to generation. If genes were jumping around, it seemed likely that this would extensively damage the transmission of genetic information and mess things up. There could be no stable heredity. McClintock's hypothesis seemed highly unlikely to be true.

The genetic element that McClintock had discovered is now called a transposon or jumping gene. Transposons are now recognized as a standard genetic component, present in all organisms and the basis for many common biological processes. But at the time her work was not much appreciated and was often treated with hostility. She finally gave up trying to promote it. She stopped publishing in 1953 and later stopped giving lectures on her findings. Yet there were people who were intrigued by McClintock's research and wanted to understand her findings better. In the 1970s I was a biology graduate student at Princeton University. One of my professors, Bruce Alberts, who went on to become the president of the National Academy of Sciences and editor-in-chief of the prestigious journal *Science*, among other major scientific accomplishments, became interested in McClintock's work and invited her to travel from Cold Spring Harbor to Princeton to deliver a lecture on it.

McClintock was not interested. She said that when she gave lectures people did not understand what she was saying. It was a waste of her time and the time of the scientists in the audience. But Alberts was truly interested and asked what it would take for her to explain herself fully. She estimated that it would take five lectures, and thus he invited her to come to Princeton for a week to present five lectures on five successive days.

Princeton University had a full schedule of lectures by famous scholars: Nobel laureates as well as Pulitzer Prize, Booker Prize, and Fields Medal recipients. In attending these lectures, I quickly learned that the physical appearance of the lecturer was unrelated to their intellectual stature. Some speakers were well-dressed and impressive in their appearance while others were poorly groomed, unkempt, and disheveled. Barbara McClintock was a tiny, unimposing older woman, perhaps five feet tall. But this little woman listened to everything intently, meticulously taking it all in. Her eyes were in constant motion, always appraising her surroundings and those around her. And behind those eyes there was an extraordinarily active and inquisitive mind.

Barbara McClintock was probably the most intellectually honest person I have ever had the good fortune to meet. When scientists lecture on their work, they commonly first present their hypothesis and then present their data in support of their hypothesis. But McClintock appeared to think that this approach was not completely honest, since stating the hypothesis up front would tend to direct the thoughts of her listeners away from other possible explanations of the experimental results. Her approach was instead to present her data and trust that the power of her experimental design and execution would lead the listeners to the same ineluctable conclusion that she had arrived at. Her approach was arguably fairer to the scientific process, but it made her presentations much, much harder to understand.

In classical genetics studies, scientists work by mating selected individuals, and then they carefully count and record the resulting offspring. In McClintock's Princeton seminars, she described the crosses she had made, and then the counts of the appearance of color variations in corn kernels generated from her crosses. That was it. We saw crosses, the resulting ears of corn, kernel counts, then tables listing the different types of kernels with counts of the numbers of each kernel she had observed, followed by more genetic crosses, more ears of corn, and more kernel counts, more tables, for an hour a day over five days.

I understood nothing. I felt like I was drowning in data the entire time. I wondered, did my classmates understand what McClintock was saying while I did not? Did the faculty understand her presentation? I worried that perhaps I was just too stupid to comprehend the great scientist. But thirty years later I saw a scientific article by Marc Kirschner, one of the Princeton faculty in attendance, who later went on to become the first chairman of the department of systems biology at Harvard Medical School. In this paper Kirschner included comments on McClintock's Princeton seminars, writing, "I remember Bruce [Alberts] inviting Barbara McClintock to lecture on cytogenetics. I understood little of what she said, yet the impact of her enthusiasm and tireless intellect is still with me." Kirschner's paper certainly made me feel a lot better about my own level of intelligence.

All of McClintock's evidence for gene movement was by inference. There was no way for her to show directly that her genes had actually moved. McClintock's findings on transposable genetic elements were published in 1950. James Watson and Francis Crick's classic paper, which showed for the first time that genes were made of DNA, appeared in 1953, three years after the publication of McClintock's jumping gene paper and two years after her follow-up Cold Spring Harbor talk. In fact, when she was performing her studies, most scientists at the time thought genes were composed of proteins. It was thought that the structure of DNA was far too simple to encode complex genetic information. Not knowing what genes were made of, it was impossible for McClintock to track her genetic elements physically or chemically.

But by the 1970s new experiments were being performed that led to the independent discovery of transposons. Penicillin, the first antibiotic, had been introduced into clinical use in the spring of 1942. Resistance to penicillin started to appear by 1944. Other antibiotics, erythromycin, tetracycline, streptomycin, etc., were discovered during the next decade, and antibiotic resistance soon appeared after the clinical introduction of each new antibiotic. Even more concerning, resistance in many cases seemed to spread like wildfire among the pathogenic bacteria that the antibiotic was being used to treat. This was a medical crisis, and a large number of research laboratories were funded to study antibiotic resistance.

From a research perspective it was fortunate that all of this was happening in bacteria, which were the best targets for the application of the newly developing DNA research methods. The genome is a complete sequence of DNA in a cell, measured in base-pairs of DNA. Bacterial cells have on average about 4 mega base-pairs of DNA (about 4 million base-pairs). Humans have 3,100 mega base-pairs of DNA (about 3 billion base-pairs, about a thousandfold more than bacteria). And the human genome is roughly about the same size as the corn genome, which has about 2,500 mega base-pairs. From the perspective of molecular biology, bacteria were going to be a thousand times easier to study than corn.

During the intense scientific investigation effort on antibiotic resistance from the 1950s through the 1980s, it was straightforward to imagine how antibiotic use would select for antibiotic resistance. But why was resistance spreading so quickly? Bacteria had been believed to be asexual. A bacterium could thus donate its antibiotic resistance gene to its progeny but would not be expected to donate a copy of its antibiotic resistance gene to other bacteria and thus rapidly spread the gene around the population.

Studies soon provided evidence for mechanisms for genetic exchange in bacteria. These studies showed that antibiotic resistance genes were often carried on transposable genetic elements, transposons, bacterial jumping genes. Transposons were a perfect vehicle to spread antibiotic resistance. The antibiotic resistance genes could ride on the transposons as they jumped from pathogenic strain to pathogenic strain. This was exactly what Barbara McClintock had been saying, but, unlike in corn, and following the development of scientific methods to analyze DNA, it was relatively simple to provide evidence for this in bacteria. The pieces of DNA carrying the antibiotic resistance genes could be analyzed chemically. It was therefore possible to show directly that the piece of DNA carrying the resistance gene moved from one place to another. This result provided direct evidence that could quickly convince doubting scientists about the validity of the jumping gene model.

Obviously, the discovery of how antibiotic resistance spread was delayed for years, if not decades, because the scientists working on the problem had to rediscover independently what McClintock had found many years earlier. Today, knowledge of how transposons behave is of extreme importance, enabling public health officials to better predict how drug resistance or virulence factors in an infectious disease can disseminate within a population.

McClintock had done something amazing, especially considering the methods that had been available to her. Many later scientists worked to make sure that the pioneer was credited with the discovery, even if her peers at the time had not understood or appreciated her work. One of these scientists was Nancy Kleckner, who was a professor at Harvard during the late 1970s. Kleckner worked with Tn10, a transposon that determines resistance to the antibiotic tetracycline. She and her colleagues were great supporters of McClintock's work, explaining to others how McClintock had found the same thing they had found but had done so decades earlier while working with limited tools and using a much more difficult system.

In the early 1980s the Ac and Ds genetic elements that McClintock had studied thirty years previously were molecularly cloned and sequenced. It turned out that the Ac element was a small typical garden-variety transposon, a simple transposable element with a totally conventional structure. It was pretty much the same thing that the antibiotic resistance researchers had been finding. Ds elements were found to be derivatives of Ac but with parts of the Ac sequence missing. Barbara McClintock was awarded the Nobel Prize in 1983 at the age of eighty-one for her work on Ac and Ds.

With the confirmation provided by subsequent research, it was now clear that McClintock's transposons were real and jumped from place to place. But how could this be reconciled with the very reasonable idea that if genes were jumping around, it would totally mess up inheritance? This conundrum was addressed by an accidental line of investigation.

As described in the preceding pages, in the early twentieth century Thomas Hunt Morgan developed methods to carry out genetic experiments using the fruit fly, *Drosophila melanogaster*. With time this model organism became increasingly popular with geneticists. The fruit fly system was easy and cheap to use and, as more and more laboratories began working with the tiny flies, it opened up numerous opportunities for research collaborations. Soon genetic studies using fruit flies were being performed in hundreds of laboratories around the world.

Work in all of these laboratories used flies that were direct descendants of the handful of fruit flies that Morgan had collected from a fruit vendor's cart on Broadway in New York City in 1908. But in the 1960s several laboratories tried something new. They designed experiments in which they mated their laboratory flies with wild flies they had freshly collected from nature.

When these scientists analyzed their results they were shocked to see that all hell had broken loose. The progeny offspring from these matings had all sorts of terrible defects: mutations, sterility, defective gene transmission, and chromosome damage. The scientists who made these observations quickly set aside their original scientific goals to try to find out what in the world was going on.

It took almost twenty years to figure out, but they eventually determined that all the problems they had been seeing were due to the appearance of a new transposon. During the more than fifty years from the time when Morgan had first collected his specimens, a new transposon had somehow gotten introduced into wild fly strains. The laboratory strains, effectively in quarantine for more than half a century, had never been exposed. But then they suddenly were because of the new experimental design. Scientists named the new transposon the *P element*.

The first exposure to the P element had obviously produced significant damage. But then things seemed to settle down. As destructive as the P element was at first, it also must have done good things, improving the fitness of the wild fruit flies. Rather than this transposon killing off the fruit flies, now almost all of the flies in the wild carried it and were doing just fine.

When Barbara McClintock started her experiments. almost all scientists thought that transposons could not possibly exist. They thought that transposons would be incompatible with life. Later on, as experiment after experiment confirmed the existence of transposons, many scientists began to call transposons "selfish DNA" or "junk DNA," an idea famously described in a 1976 book by Richard Dawkins titled *The Selfish Gene*. These scientists proposed that transposons were DNA parasites, getting a free DNA replication ride from their hosts but doing little or nothing to help the host in the process.

Now though, because of this unintentional fruit fly experiment, that idea had to be modified. Transposons appeared not as parasites so much as some sort of a critical part of life, producing mutations that provide the raw material for evolution. Some of these new mutations turn out to be highly beneficial, thus improving the odds of survival for the transposon-carrying organisms.

One striking example of what happens was illustrated when a fruit fly P element set down on a specific gene. After the P element landed, the lifespan of the flies was extended by approximately 35 percent. The flies carrying this P element insertion also showed increased resistance to starvation, damaging high temperatures, and other environmental stresses. Scientists named this P element landing site the *methuselah gene*.

Taking a longer-term perspective, it is clear that during the course of evolution host organisms have co-opted transposon functions to serve their own needs. Transposons possess an exquisitely powerful apparatus that is able to cut and paste DNA and move DNA sequences around. Hosts have taken advantage of these transposon mechanisms in several ways. Transposons are highly regulated, sometimes jumping and at other times remaining in place. In some cases, hosts have adapted transposon control mechanisms to regulate their own genetic processes.

Higher cells require a highly specialized system to replicate the ends of their chromosomes. It appears that aspects of this system were recruited in ancient times from transposons and then adapted to the needs of the host, allowing them to accurately replicate their chromosomes generation after generation. And it appears that the exquisite ability of transposons to cut and paste DNA has been taken over by host cells and then repurposed to remove mutations produced by X-rays and chemical mutagens. Thus, transposons are at once agents of mutagenesis and have served as the raw material to produce systems that oppose mutagenesis.

Finally, with the advent of genomic DNA sequencing, it became possible to know exactly how many transposons are present in many organisms, including humans. Surprisingly, it was revealed that 45 percent of the human genome is composed of transposons or transposon fragments, almost half. So instead of being a hindrance to the propagation of life, transposons appear instead to be a fundamental aspect of life. It took a long time, but in the end McClintock's ideas moved from scientific heresy to mainstream science.

McClintock, a tiny woman with an enormous intellect, was a gracious person, a great mentor, and a kind and supportive individual who gladly and selflessly invested her own time and energy to help others, particularly other women scientists. Visitors to Cold Spring Harbor Laboratory (including me—I spent the summer of 1980 working there) would often see her walking the grounds discussing the latest genetic research with other scientists. She was always eager to learn something new or to help someone with a research problem. She never married and devoted her entire life to the pursuit of scientific knowledge. Her discoveries were counterintuitive and challenged by other scientists, but in the end her high scientific standards, powerful deductive reasoning ability, and pursuit of important problems led to her making some of the most important discoveries of the twentieth century.

4 Galileo Galilei
Astronomer who determined the structure of our solar system

Galileo Galilei was born in Pisa, Italy, in 1564. His parents were Vincenzo Galilei, a lutenist, composer, and music theorist, and Gulia Ammannati, a noblewoman from a prosperous and influential Italian family. One of his mother's ancestors was Iacopo Ammannati, who served as the secretary of Pope Pius II during the fifteenth century. Galileo was named after another famous ancestor, Galileo Bonaiuti, who was a prominent fifteenth-century physician, university scholar, and politician.

Galileo had three siblings, a brother and two sisters. His younger brother, Michelangelo, also a lutenist, was a significant financial drain on the family. Michelangelo did not contribute his fair share to his sisters' dowries and periodically borrowed money from Galileo to support his musical endeavors. These financial burdens encouraged Galileo to pursue a career that would provide him with money, so he rejected vocations his father encouraged, such as music or the priesthood.

When Galileo was eight, his family moved to Florence, where he was sent to study at the nearby Vallombrosa Abbey. In 1580 Galileo enrolled at the University of Pisa for a medical degree. Despite the allure of the good income he could earn as a physician, he found himself pursuing topics of scientific interest with an eye to turning his new knowledge into profitable inventions. As an early example, while he was in medical school Galileo showed that the length of a pendulum determined the frequency of its motion and that the frequency was independent of the distance the pendulum traveled while it swung. This was an important scientific finding, but it took another one hundred years for Galileo's observation to be exploited by Dutch physicist Christiaan Huygens to create a clock.

After attending a lecture on geometry by chance, Galileo talked his reluctant father into letting him study mathematics and natural philosophy rather than medicine. In 1586 he published a book on the design of a hydrostatic balance he had invented, and in 1593 he invented the thermoscope, an early thermometer. Galileo also studied disegno, a method for creating new designs and portraying them on paper.

In 1588 Galileo was appointed as an instructor at the Accademia delle Arti del Disegno in Florence, where he pursued both artistic and scientific studies. A year later, he was appointed as chairman of the mathematics department in Pisa, and in 1592 he moved to the University of Padua, where he taught mechanics and astronomy until 1610. It was in Padua that Galileo made the important discoveries for which he is best known today.

The Earth Is Not Flat

It is commonly taught that everyone in the fifteenth century thought the earth was flat and that Columbus set out sailing westward from Spain to find a new route to the Orient and to prove that the world is round. But this is not true. All learned people for well over a thousand years prior to Galileo's birth knew that the earth is round. The evidence for this came from a few simple observations.

Sailors returning to shore observed that they were able to see the peaks of mountains before they were able to see the mountain base. If they were traveling on a level surface, one would expect the entire mountain to come into view at once. Similarly, to observers on shore when a ship sails off toward the horizon, it does not just get smaller and smaller until it is no longer visible. Instead, the hull seems to sink below the horizon first, then the mast. And when ships return from sea, the sequence is reversed. You first see the mast, and then the hull appears to rise over the horizon. Since it is hard to imagine anything flatter than a calm sea, the explanation for these observations must be that the earth is round.

Secondly, people see different sets of stars in the night sky depending upon where they are. Ancient Greek travelers reported significant differences between the night sky seen in the Nile Delta and in Crimea. If the earth was flat, everyone would see the same stars at any given time. They do not.

Eratosthenes, a Greek philosopher who lived circa 200 BC, employed observations of the sun to estimate the circumference of the earth. On a day when the sun was directly overhead in one Egyptian city, it did not rise as high in a distant city. Knowing the height of the sun in the two places and the distance between the two cities, trigonometric calculations allowed Eratosthenes to determine a value for the circumference of the earth that is not very different from what we know to be true today.

But the idea that the earth is round was not without its issues. Some people in the northern hemisphere worried that people in the southern hemisphere would fall off and vice versa. Despite any and all discomfort with the idea, Ferdinand Magellan led a Spanish expedition that from 1519 to 1522 circumnavigated the earth, putting the issue to rest for all time.

Is the Earth the Center of the Universe?

That the earth is round was considered to be settled science during Galileo's lifetime. The major question facing thinkers of Galileo's day was whether the earth was the center of the universe (a theory known as geocentrism) or, instead, whether the earth revolved around the sun (heliocentrism). The prevailing theory when Galileo was born was geocentrism; that the sun, moon, stars, and planets all revolved around the earth. The geocentric model had been established in antiquity based upon two commonsense observations.

First, from everywhere on earth the sun appears to revolve around earth once per day. The moon and the planets have their own different motions, but they also appear to revolve around earth once per day. And the stars appear to be fixed on a distant celestial sphere that also revolves around the earth once each day on an axis through the north and south poles of the earth. Second, the earth seems to be unmoving. It feels solid, stable, and stationary, terra firma as the Romans said.

Arguments against this would seem to conflict with everyday observations. But it was long known that there were small observed inconsistencies with the simple model of the stars and planets just revolving around the earth. For example, the seasons are of unequal length. In the northern hemisphere, autumn is about five days shorter than spring. (We now know that for the seasons to be of equal length the earth's orbit would have to be perfectly circular, and that the earth's orbit is actually an ellipse.)

In addition, planets sometimes slow down, stop, move backward in retrograde motion, and then again reverse to resume normal, forward motion. Very tricky modeling was needed to fit these observations with geocentrism.

Claudius Ptolemy was a mathematician, astronomer, geographer, and astrologer who lived in Alexandria during the second century AD. He developed a detailed model to explain these aberrant planetary observations while retaining the earth at the center of the universe. Ptolemy's model would stand for almost 1,500 years.

In the Ptolemaic system the earth is placed away from the center of rotation of the rest of the universe. This explains different season lengths. In addition, to explain periodic retrograde planetary motion, Ptolemy said that each planet is moved by an arrangement of two spheres, one called its deferent and the other its epicycle. The deferent sphere has the earth at (or more precisely near) its center. The epicycle is a sphere that rides around on the deferent sphere and allows the planets to change direction.

Aristotle argued that celestial bodies maintained constant circular motion. It was a clear, simple, logical, and satisfying idea. But by Ptolemy's time astronomers had made observations that conflicted with Aristotle's constant circular motion proposition. To reconcile his model with Aristotle's thinking, Ptolemy came up with the idea of the equant, which resolved the conflict. The equant is a geometrical point at which the planet's epicycle always appears to move at uniform speed. The earth is not located at the equant, so we do not observe constant circular motion. For us the planets' motions appear to be nonuniform in speed. Therefore, Aristotle was right. It is just that the earth is in the wrong place for us to see it.

Ptolemy's arguments were tortured. But his system came to be widely accepted in Western thought. Despite the fact that it was clearly cumbersome and unwieldy, it did accurately predict the known various celestial motions in a way that was consistent with geocentrism, and everyone knew that geocentrism was correct. But sixteenth-century science came to embrace a different idea that better explained astronomical observations: the revolutionary idea that the sun and not the earth was at the center of the universe.

Copernicus

Nicolaus Copernicus lived during in the late fifteenth and early sixteenth centuries. He was a scholar known for his investigations in multiple fields including theology, mathematics, medicine, and astronomy. In the year 1514 Copernicus wrote up the *Commentariolus* (Little Commentary), a brief outline of his idea that the earth was not at the center of the universe but rather that it revolved around the sun.

In 1532 Copernicus completed a detailed description of this idea, which he titled *De revolutionibus orbium coelestium* (*On the Revolutions of the Heavenly Spheres*), which was first published in Nuremberg in 1543. Copernicus's model incorporated many novel concepts, including the idea that the earth is one of several planets revolving around a stationary sun in a determined order, that the earth has three motions (daily rotation, annual revolution, and annual tilting of its axis), that retrograde motion of the planets is due to the earth's motion taken together with the planets' motions, and that the distance from the earth to the sun is small compared to the distance from the sun to the stars.

This last idea was highly important because a major argument against the earth revolving around the sun was the lack of parallax in viewing the stars. Parallax is the displacement or difference in the apparent position of an object when viewed along two different lines of sight. (For example, the driver of a car will look at the speedometer directly in front of her and see the needle at exactly 60 miles per hour. But the passenger looks at the speedometer from a different angle, and he incorrectly sees a different speed, a common cause of marital disharmony.)

If the earth revolved around the sun, one would expect that there should be significant parallax observing a star when the earth is on one side of the sun compared with observing the same star when the earth is on the opposite side of the sun. Parallax was not observed in ancient times because ancient astronomers' instruments were too imprecise to detect it. This observation was mistakenly taken as a proof that the earth cannot revolve around the sun. But Copernicus argued that if the stars were extremely distant from the earth parallax would be far too small to measure. We know this to be true today.

Publication of *De revolutionibus orbium coelestium* was an extremely bold step, and it isn't surprising that the book was first published more than ten years after it was written. Copernicus's hypothesis contradicted the Old Testament account of the Sun's movement around the Earth (Joshua 10:12–13). Thus, for the Catholic Church, *De revolutionibus orbium coelestium* was heresy.

The publication contained an unauthorized, anonymous preface saying that Copernicus had written his manuscript as a mathematical hypothesis, not as an account that contained truth or even probability. It was later determined that this preface had been added by a Lutheran preacher named Andreas Osiander, who lived in Nuremberg when the first edition was printed there. Presumably, Osiander was a friend of Copernicus who was trying to help him stay out of trouble with the Church. In addition, Copernicus dedicated the book to Pope Paul III and said that his purpose in writing the book was to increase the accuracy of astronomical predictions to permit the Church to develop a more accurate religious calendar. Reforming the Julian Calendar, inherited from Roman times, was a major goal of the Church. Thus, Copernicus argued that his intention was to support the Church rather than challenge it. In any event, at the time there were no data to conclusively show whether geocentrism or heliocentrism was the correct model.

A New Instrument Advances Scientific Knowledge

The telescope was invented in the Netherlands in 1608 by the eyeglass maker Hans Lippershy. News of the new invention rapidly spread across Europe. Lippershy's invention was basically a spyglass, a low-power refractor telescope that amounted to half a pair of binoculars (but without the erecting prism). His design had two lenses at the opposite ends of a hollow tube: an objective lens that gathers a large amount of light and brings it to a focus to create an image, and an eyepiece lens that magnifies the image. This design met the dual challenges of seeing faraway objects: their dimness and their small size.

When Galileo Galilei heard about Lippershy's invention, he immediately sought to improve it. Galileo had no diagrams to work from, so his method for improvement was trial and error, working to optimize both the lenses and their placement. Lippershy's instrument had achieved a magnification of only two- to three-fold. Galileo's first-generation telescope provided eight-fold magnification, and later versions increased this to twenty-three-fold. The last telescope Galileo made was the first instrument to actually be called a telescope, a name invented by the Greek poet and theologian Giovanni Demisiani. The word *telescope* comes from the Greek words *tele*, meaning "far," and *skopein*, meaning "to look or see." Galileo's telescopes had a narrow angle of view, but that was not a disadvantage for his intended use: to look at tiny, faraway objects in the night sky.

On March 13, 1610, Galileo published the observations he had made with his new telescope in a pamphlet titled *Sidereus Nuncius, The Starry Messenger*. In it he described four new findings, observations that were astonishing to people at the time. He saw that the uneven appearance of the moon was not simply due to differences in pigmentation but rather due to the fact that surface was composed of mountains and broad flat plains, "uneven, rough, full of cavities and prominences." He saw hundreds of new stars in constellations and in the Milky Way that were invisible to the naked eye.

He then made two observations that had profound implications in light of Copernicus's theory. When Galileo looked at the stars, even with his telescope, he saw pinpoints of light. But when he looked at Jupiter, he saw a disk. And there were four "stars" arranged in a line near to the left and right of the disk of Jupiter. He observed these four "stars" night after night. Sometimes two of them were on one side and two were on the other side of Jupiter. Sometimes three were on one side and only one was on the other side of Jupiter. And sometimes one of the "stars" disappeared only to reappear at a later time on the other side of Jupiter.

Galileo came to the conclusion that these objects were not stars but rather moons of Jupiter and that these moons revolved around Jupiter. The idea that the earth was at the center of the universe and that everything revolved around it could not be true! Here was a clear exception and one that could be easily verified by anyone who had a telescope. The moons of Jupiter revolved around Jupiter, not the earth. So, perhaps other astronomical objects also did not revolve around the earth.

Galileo then looked at the planet Venus and made a crucial observation. Unlike the stars and similar to Jupiter, Venus was a disk. But unlike Jupiter, it did not consistently appear as a round disk. Over time Venus changed its appearance and had phases like the moon. Sometimes it was an almost full disk, called a gibbous phase. Other times it was a half circle. At still other times it appeared as a crescent.

Venus is never very far from the sun in our sky. It is either a morning or an evening "star," rising just before or setting just after the sun. According to Ptolemy, Venus could only be in a crescent phase, because it moves around on an epicycle that is never far from the sun. For Venus to appear as a gibbous object, Venus and the earth had to be on the opposite sides of the sun, something not permitted in the Ptolemy model. But a gibbous phase would be predicted if Venus went around the sun together with the earth and not around the earth.

This experimental evidence provided a definitive test of the two models, geocentrism and heliocentrism. Previously, both the geocentric and heliocentric models were able to describe all available data, although with considerable force-fitting on the part of the geocentric model. Now there was conclusive evidence that the geocentric model was wrong. Galileo published his observations of phases of Venus together with his studies of sunspots in 1613 in a pamphlet titled *Istoria e Dimostrazioni intorno alle Macchie Solari* (Letters on Sunspots). And that is when Galileo's troubles started.

Problems with the Church

The Jesuit order was founded in 1534 in order to pursue both faith and learning. Therefore within a year or two of Galileo's publication of *The Starry Messenger*, Jesuit astronomers obtained their own telescopes and repeated his observations. Not surprisingly, they saw the same things. Also not surprisingly, their interpretations of what they saw were totally different.

In 1611 Galileo visited the Collegium Romanum in Rome to meet with the Jesuit astronomers who had repeated his observations. The official position of the Jesuits was to defend the idea that the earth is the center of the universe. In a letter to fellow astronomer Johannes Kepler, Galileo complained that some of the Jesuits who opposed his ideas had refused to even look through a telescope. It was a seventeenth-century version of the 2021 motion picture *Don't Look Up*.

In the years following his initial publications, Galileo's undiplomatic and contentious style made him many enemies. Tommaso Caccini, a Dominican friar, undertook the first damaging attack on Galileo in 1614. Then in early 1615 Caccini's fellow Dominicans brought Galileo's writings to the attention of the Inquisition, denouncing Galileo for his defense of Copernicus's theories and various other alleged heresies. On February 19, 1616, the Inquisition asked a commission of theologians, known as Qualifiers, about the propositions of the heliocentric, or sun-centered, view of the universe.

On February 24 the Qualifiers delivered their unanimous report. The proposition that the sun is stationary at the center of the universe is "foolish and absurd in philosophy, and formally heretical since it explicitly contradicts in many places the sense of Holy Scripture" and the proposition that the earth moves and is not at the center of the universe "receives the same judgment in philosophy; and . . . in regard to theological truth it is at least erroneous in faith."

Galileo was ordered to abandon the Copernican opinions and warned that resistance would result in stronger action. A 1616 Inquisition injunction against Galileo ordered him "to abstain completely from teaching or defending this doctrine and opinion or from discussing it . . . to abandon completely . . . the opinion that the sun stands still at the center of the world and the Earth moves, and henceforth not to hold, teach, or defend it in any way whatever, either orally or in writing."

On March 5 of that year the papal Congregation of the Index banned all books advocating the Copernican system, which it called "the false Pythagorean doctrine, altogether contrary to Holy Scripture." Galileo's works advocating Copernicanism were banned as were Kepler's works on heliocentrism. In particular, Copernicus's *De revolutionibus orbium coelestium* was banned and remained on the index of banned books until 1758.

While asserting as fact that the sun is the center of the universe was banned, there appears to have been some latitude for discussing geocentrism and heliocentrism simply as opposing philosophical ideas. In 1632 Galileo published *Dialogo sopra i due massimi sistemi del mondo* (Dialogue Concerning the Two Chief World Systems). The book presented a series of fictional discussions about geocentrism and heliocentrism over a span of four days among two philosophers and a layman. It summarized previously published arguments in favor of and against each possibility and attributed them to the different characters in the book. But Galileo went too far in presenting these arguments. Any careful reading made it clear which system he believed was clearly correct and which was obviously wrong. And the character supporting geocentrism comes off looking like a fool.

The book was published in Florence under a formal license from the Inquisition. But once it appeared it kicked up a violent storm of Church disapproval. In 1633 Galileo was ordered to stand trial on suspicion of heresy "for holding as true the false doctrine taught by some that the sun is the center of the world." Galileo was interrogated and threatened with physical torture. A panel of theologians determined that the Dialogue Concerning the Two Chief World Systems taught the Copernican theory. Galileo was sentenced to formal imprisonment at the pleasure of the Inquisition. But on the following day this was commuted to house arrest. Galileo remained under house arrest in the Arcentri Hill above Florence for the next eight years until his death in 1642.

Galileo's Legacy

According to legend, after his conviction and sentencing Galileo defiantly commented, "Eppur si muove" (And yet it moves). For years Galileo's standing was questioned at every turn. In March 1641 an Inquisitor required the author of a book about astronomy to change the words "most distinguished Galileo" to "Galileo, man of noted name."

But with time things began to change. In 1657 an early scientific society, the Accademia del Cimento, was founded in Florence. The society was based on the principles of experimentation, avoidance of speculation, creation and use of laboratory instruments, and adherence to standards of measurement. Its motto was *provando e riprovando* (proving and disproving).

The position of the Church began to change as well, though more gradually. In 1758 the Catholic Church dropped the general prohibition of books advocating heliocentrism, but uncensored versions of Copernicus's *De revolutionibus orbium coelestium* and Galileo's *Dialogo sopra i due massimi sistemi del mondo* remained banned. The 1633 Inquisition judgment against Galileo also remained in force.

Following an appeal to Pope Pius VII in 1820, the Church's ban on books advocating heliocentrism was further lifted and the books by Copernicus and Galileo were omitted from the 1835 edition of the Index of Forbidden Books. In 1979 Pope John Paul II encouraged that "theologians, scholars and historians, animated by a spirit of sincere collaboration, will study the Galileo case more deeply and in loyal recognition of wrongs, from whatever side they come."

A Pontifical Interdisciplinary Study Commission was constituted in 1981 to study the Galileo case, but it did not agree to take any action. And then on February 15, 1990, Cardinal Ratzinger, who later became Pope Benedict XVI, quoted the radical philosopher Paul Feyerabend, saying, "The Church at the time of Galileo kept much more closely to reason than did Galileo himself, and she took into consideration the ethical and social consequences of Galileo's teaching too. Her verdict against Galileo was rational and just, and the revision of this verdict can be justified only on the grounds of what is politically opportune." And this is pretty much where the Catholic Church stands today.

In his time Galileo's ideas were opposed by members of the astronomy community with the distinct concomitant problem that his writings went against the Church's vested opinions on astronomy. Absent the Church, Galileo together with like-minded astronomers might have been able to unite the astronomy community behind his ideas. But because of the doctrinal interests of the Church, Galileo ended up under house arrest and had his writings banned for well over a century.

Not everyone agrees with the Church, especially scientists.

Today the scientific community regards Galileo to be the father of observational astronomy, modern physics, the scientific method, and modern science. But for some people, four hundred years of supportive experimental evidence has not been enough to change their thinking.

5 | Ignaz Semmelweis
Pioneer of hospital infection control

During the COVID-19 pandemic we all became accustomed to seeing images of medical workers suited up in elaborate personal protective equipment. And it is common knowledge that behind the scenes at every hospital, a complex array of sanitization and decontamination protocols were, and as a general practice still are, being followed. This was being done to protect medical workers from SARS-CoV-2, the virus that causes COVID-19 disease, as well as to prevent medical workers from exposing patients at the hospital to the virus and, when they returned home, to protect their families and friends. Most everyone saw this as pretty reasonable and logical. But two hundred years ago, that was not the case.

Up until the late nineteenth century, physicians did not accept what we now call the *germ theory of disease* to be true. It took half a century of work by a group of physicians and medical scientists to convince the medical establishment that many of the most important diseases were caused by pathogenic microorganisms, what we commonly call *germs*. This group of researchers showed that in many diseases, tiny organisms invisible to the naked eye invaded human bodies, where they grew and reproduced and by so doing caused disease.

Childbed fever, also called puerperal fever, is caused by the bacterial infection of the female reproductive tract following childbirth. In developed countries today the occurrence of childbed fever is low: about 1 to 2 percent following vaginal delivery and perhaps as much as 10 percent in complicated deliveries. And today such infections are easily treated with antibiotics.

But in ancient times childbed fever was a common cause of death among childbearing women. It equally affected wealthy women and poor women, famous women and obscure women. Lucrezia Borgia, the famous Spanish-Italian noblewoman of the House of Borgia, was the daughter of Pope Alexander VI and Vannozza dei Cattanei. She reigned as the governor of Spoleto, a position traditionally held only by cardinals. During her lifetime, Borgia's social standing was further advanced by three advantageous marriages to prominent noblemen. Borgia was a femme fatale of her era.

In 1519, at the age of thirty-nine, Borgia gave birth to a female child whom she named Isabella Maria. Borgia had become very weak during the pregnancy, and she fell seriously ill immediately after the birth with a high fever. Following a hopeful but temporary recovery, she died two days after the birth and was buried in the convent of Corpus Domini. The cause of her death was childbed fever. At the time no one knew what caused childbed fever, and nothing could be done to prevent it or treat the patient.

Not much about childbed fever changed for the following three centuries. What we know about the disease today was well beyond the comprehension of physicians until fairly recent times. How medical science came to this knowledge is the story of a nineteenth-century physician named Ignaz Semmelweis.

Early Career

Ignaz Semmelweis was born in 1818 in Budapest to József Semmelweis and Teréz Müller, a prosperous grocer family of ethnic Germanic ancestry. Semmelweis began studying law at the University of Vienna in 1837 but for unknown reasons switched to medicine the following year. He was awarded his doctor of medicine degree in 1844 and decided to specialize in obstetrics.

Semmelweis was appointed assistant to Professor Johann Klein in the First Obstetrical Clinic of the Vienna General Hospital in 1846, a position comparable to the current American position of chief resident. At that time most babies were born at home. However, hospital maternity wards were established in the early nineteenth century to counter the infanticide of illegitimate children and mainly served underprivileged women and prostitutes. In return for the free services, these women served as subjects for the training of doctors and midwives.

During the 1840s Vienna General Hospital had two maternity clinics. The first division clinic was established for the training of physicians, and the second division clinic for the training of midwives. The first division clinic had a childbed fever maternal mortality rate of approximately 18 percent, while the second division clinic had a rate of approximately 2 percent. The two clinics admitted on alternate days, and women did whatever they could to get into the second division clinic, which was known to have the lower mortality rate, often delaying admission and not uncommonly literally giving birth on the street outside the hospital in order to gain admission to the better clinic.

Semmelweis was puzzled about why women who had given birth on the street and were then admitted to the second division clinic had a lower rate of childbed fever than the women who were tended to in the first division clinic. He later wrote in his master-work, *The Etiology, Concept, and Prophylaxis of Childbed Fever*, "To me, it appeared logical that patients who experienced street births would become ill at least as frequently as those who delivered in the clinic. . . . What protected those who delivered outside the clinic from these destructive unknown endemic influences?"

In his day, childbed fever was thought to have myriad causes, including outlandish ones such as the mothers' embarrassment at being examined by male doctors. To achieve a more reasoned understanding, Semmelweis began the meticulous process of examining every possible difference between the first and second division clinics, thus embarking on a lifelong campaign to deter-mine which clinical practices were critical to producing higher or lower rates of childbed fever.

Semmelweis created a list of all the differences between the two clinics and then set about ruling out the irrelevant ones. For example, in the second division clinic women gave birth on their sides. In the first division clinic, women gave birth on their backs. So Semmelweis had women in the first division clinic give birth on their sides. No effect.

In the first division clinic, whenever a woman died of child-bed fever a priest would walk around the hospital ward past the women's beds with an attendant ringing a bell. This was not done in the second division clinic. Semmelweis theorized that the priest and the bell ringing so terrified the women after birth that they developed a fever, got sick, and died. He had the priest change his route and ditch the bell. No effect. Semmelweis also excluded overcrowding as a cause, since the second division clinic was more crowded than the first division clinic and still had lower mortality. Similarly, he excluded any effect that the weather might have as a possible cause, since the weather for the two nearly adjacent clinics was identical.

Another difference between the two clinics was that patients in the second division clinic were attended to by midwives. In the first division clinic patients were attended to by medical students. These medical students, in addition to treating obstetric patients, were required to perform autopsies on patients who had died in the hospital.

Through the serendipity of a tragic accident, Semmelweis made a breakthrough in 1847. His colleague and good friend, Jakob Kolletschka, had been performing an autopsy on a diseased childbed fever patient one day and was accidentally poked with a student's scalpel. Kolletschka died shortly thereafter. His autopsy showed pathology that was identical to the pathology of the childbed fever patients. Semmelweis proposed that childbed fever and death from cadaveric contamination were the same disease.

He hypothesized that the medical students had "cadaverous particles" on their hands from their work in the autopsy room and carried these cadaverous particles to the maternity patients they examined upon leaving the autopsy room. No one in the second division clinic performed autopsies or had any contact with cadavers, thus suggesting a cause-and-effect relationship in the incidence of childbirth fever between conducting autopsies on women who had died from childbirth fever and then attending to living and breathing obstetric patients who were waiting to give birth. In Semmelweis's hypothesis, cadaverous particles carried by physicians caused childbed fever just as particles on the student's scalpel had caused his friend's death.

What were these "cadaverous particles"? These events transpired a half-century before the acceptance of the germ theory of disease, so in his thinking Semmelweis was probably influenced by the miasma theory of infectious disease that prevailed at the time. The miasma theory was based on the observation that diseases and epidemics had a much higher occurrence in poor, malodorous slum neighborhoods than in prosperous neighborhoods. One clear difference between such neighborhoods was the foul smell of the slums. This led physicians to develop the miasma theory, which stated that diseases were caused by miasma, a noxious form of "bad air," also known as night air. It was further believed that miasma emanated from rotting organic matter.

It is hard to imagine an environment with fouler smelling air than a mortuary filled with rotting corpses. Semmelweis thus proposed that miasma in the mortuary contaminated the hands of medical students performing autopsies and that they then carried this contamination to their obstetrical patients. To break the cycle of contamination and subsequent transfer to patients, Semmelweis proposed that, after performing autopsies, the medical students should wash their hands with chlorinated lime, something we call liquid bleach today. Bleach is an excellent deodorant, but crucial to the success of Semmelweis's idea is that bleach is also an excellent disinfectant.

The result from hand washing was dramatic. After only a month of hand washing, the mortality rate in the first division student clinic dropped by 90 percent from an 18 percent patient mortality to 2 percent. Half a year later the death rate approached zero.

Dissemination of Information on the
Cause and Prevention of Childbed Fever

Five years prior to Semmelweis's clinical work, the medical community on the other side of the Atlantic Ocean was also questioning the cause of childbed fever. Oliver Wendell Holmes Sr. was a nineteenth-century American physician and medical reformer practicing medicine in Boston. At one point Holmes quipped that if all contemporary medicine was tossed into the sea, "it would be all the better for mankind—and all the worse for the fishes." In 1843 he published a paper titled "The Contagiousness of Puerperal Fever" in in an obscure medical journal, *New England Quarterly Journal of Medicine and Surgery*. In the paper Holmes argued that childbed fever was transmitted from patient to patient via contact with their physicians. But no one took much note of his contention, in no small part because, even if Holmes was correct, he had proposed no clinical practice that would slow the spread of the disease. After all, physicians could not simply stop touching their patients.

Reports of Semmelweis's work began to appear toward the end of 1847. Semmelweis and his students wrote letters to the directors of several prominent maternity clinics describing their findings. Ferdinand von Hebra, an influential Viennese physician and an editor of a prestigious medical journal, was one of Semmelweis's most ardent supporters. Hebra reported Semmelweis's discovery, that the incidence of childbed fever could be dramatically reduced through handwashing with bleach, in the December 1847 and April 1848 issues of the medical journal he edited. In late 1848 one of Semmelweis's former students presented a lecture explaining his mentor's work to the Royal Medical and Surgical Society in London and subsequently published a review of his lecture in the prominent medical journal the *Lancet*. In 1849 another of Semmelweis's students published an essay on Semmelweis's work in the French medical journal periodical *Gazette Médicale de Strasbourg*. One would think everyone in the medical community would have been overjoyed to learn of Semmelweis's findings, or at least receptive to them. But things did not pan out that way.

Reactions to Semmelweis's Findings

The major obstacle to acceptance of what Semmelweis had discovered was that there was no guiding theory, based on the prevailing paradigm in medical science at the time, enabling the medical establishment to explain what was going on. Today the germ theory of disease makes it clear to everyone that the cadavers carried pathogens that produced childbed fever, and these germs were killed by the bleach. Thus, the cycle of infection was broken.

The germ theory of disease says that certain diseases are caused by infection with germs, microorganisms. Shortly before Semmelweis's death, the French scientist Louis Pasteur postulated that many common, everyday processes were likely carried out by microorganisms. He first showed that the alcohol in wine and beer is produced by single-celled yeast, *Saccharomyces cerevisiae*. He then showed that food spoilage is caused by microorganisms and developed a heating method that hindered spoilage by killing bacteria and molds present in the food, a process today called pasteurization.

Pasteur then turned his attention to pébrine and flacherie, diseases of silkworms that had reached epidemic proportions and were crippling the French silk industry. By 1870 he showed that these diseases were caused by microorganisms, the first diseases demonstrated to be caused by an infection. This result was incompatible with and directly challenged the miasma theory.

Robert Koch was a German physician and a contemporary

of Pasteur who worked in parallel with him. Koch developed a system by which he could show whether a disease was caused by an infection with a microorganism. He worked on many of the most important diseases of his era, proving that the bacterium *Bacillus anthracis* caused anthrax and that tuberculosis, at the time widely believed to be an inherited disease, was caused by the slow-growing bacterium *Mycobacterium tuberculosis*, and he provided the first evidence that cholera is caused by the bacterium *Vibrio cholera*.

The germ theory of disease didn't even exist during Semmelweis's lifetime. It became widely accepted only in the 1890s, more than twenty-five years after Semmelweis's death. With nothing but the miasma theory to guide understanding, the protocol Semmelweis had proposed had no sound theoretical basis that could convince the medical establishment. Miasma was bad air. The hands of the medical students might have been exposed to bad air, but one would expect this bad air to blow away once the medical students left the mortuary. How could the bad air turn into cadaverous particles on the medical student's hands?

Semmelweis's critics couldn't accept the notion that "minuscule and largely invisible amounts of decaying organic matter" caused every case of childbed fever. The claim that invisible and undefined entities could be the cause seemed totally absurd. One famous critic was Rudolf Virchow, a scientist who today is recognized as an influential opponent of the then widely accepted but erroneous idea of spontaneous generation, the belief that life can be produced from unliving matter. Virchow is also one of the fathers of cell theory, which states that life comes only from living matter, and the scientific dictum "Omnis cellula e cellula" (all cells come from cells). But on this one Virchow was wrong.

Lacking a firm theoretical basis, Semmelweis's ideas were rejected as being unscientific. How could contaminants in the infinitesimal amounts proposed by Semmelweis possibly cause a lethal disease? Opponents also argued that hands washed with soap and water appeared clean and yet still caused childbed fever. We now know that this argument is misleading. Soap and water is a mediocre disinfectant in contrast to chlorinated lime, which is an excellent disinfectant. Unknown at the time, Semmelweis's cadaverous particles were bacteria and would be destroyed by chlorinated lime but not so much with soap and water. Everything Semmelweis did makes good rational sense today under the germ theory.

Semmelweis's detractors had another big problem with him. The worst accusation one can make against physicians is that they are harming patients through the medical treatments they administer. But by saying that childbed fever was caused by the doctors themselves as a result of their actions while treating their patients, that was exactly what Semmelweis had done. Offended by Semmelweis's implicit accusation, the medical community reacted swiftly and severely censured the offender and his supporters.

Besides Virchow, mentioned above, three renowned professors of obstetric medicine spoke out against Semmelweis: Ede Flórián Birly, August Breisky, and Carl Edvard Marius Levy. Charles Delucena Meigs, an influential early nineteenth-century professor of obstetrics at Jefferson Medical College in Philadelphia, wrote, "How comes it then to pass, that a mortal virus or contagion should have power over a woman who is pregnant, or recently delivered, while it is innoxious for all others in the world?" In support of a fellow obstetrician who had been accused of spreading childbed fever, Meigs argued, "Did he carry it on his hands? But a gentleman's hands are clean." When Oliver Wendell Holmes Sr., then a junior resident, expressed support for Semmelweis's theory, Meigs derided Holmes as a "very young gentleman" and disparaged Holmes's clinical findings as "the jejune and fizenless dreamings of a sophomore writer."

At the same time, Semmelweis was denounced as a Jew—though in fact he came from a Christian ethnic German family. Antisemitism was rampant in mid-nineteenth century Europe and the United States. In society, it was a terrible and often calculated slur to accuse someone of being Jewish. Prominent antisemites of the era included the German composer Richard Wagner and Jacob and Wilhelm Grimm, authors of *Grimm's Fairy Tales*. Later in the century, during the Third French Republic, the Jewish artillery officer Alfred Dreyfus was falsely accused and convicted of treason by senior military officers as part of an effort to cleanse the military of Jewish influences. In the context of this sort of antisemitism, the medical establishment claimed Semmelweis's protocol was based on Jewish handwashing procedures, which were viewed as despicable and rooted in Jewish superstition.

Semmelweis had no friends in the medical establishment to support him. His appointment at the Vienna General Hospital was not renewed, and he was discharged in the spring of 1849. He petitioned to be given an alternate position, but that was denied and, in 1851, he returned to Hungary, taking a relatively insignificant, unpaid, honorary head-physician position in the obstetric ward of the Szent Rókus Hospital in Budapest, a position he held for six years until June 1857.

Semmelweis continued to fight back against his detractors. In 1858 he published a personal description of his findings in an essay titled "The Etiology of Childbed Fever." In 1860 he published a second essay, "The Difference in Opinion between Myself and the English Physicians regarding Childbed Fever," and in 1861 he published his definitive account of his findings, *Die Ätiologie, der Begriff und die Prophylaxis des Kindbettfiebers* (The Etiology, Concept, and Prophylaxis of Childbed Fever). In this work he protested against the broad-based resistance to his ideas, writing that "most medical lecture halls continue to resound with lectures on epidemic childbed fever and with discourses against my theories. . . . In published medical works my teachings are either ignored or attacked. The medical faculty at Würzburg awarded a prize to a monograph written in 1859 in which my teachings were rejected."

Although Semmelweis's procedures demonstrably worked, the medical establishment persisted in rejecting his ideas based on theoretical arguments or in reaction to what they perceived as a personal attack on established practices that he claimed were harming patients, or for both reasons.

Semmelweis knew he was right, and the stress produced by the rejection of his ideas must have been enormous. Starting in 1861, he began to exhibit a number of disturbing behaviors. He lashed out against his critics in a series of unprofessional open letters that were based on frustration, bitterness, misery, and rage rather than scientific reasoning. He accused his critics of being irresponsible murderers and called them ignoramuses. By 1865 Semmelweis's public behavior had become exasperating and embarrassing to his colleagues. He began to drink heavily and progressively abandoned his family.

Scholars have proposed two explanations for these behavioral changes. One school of thought says that Semmelweis was suffering from a medical disorder, likely Alzheimer's disease or third-stage syphilis. Third-stage syphilis was not uncommon at the time among practicing obstetricians. They examined large numbers of women, including prostitutes and impoverished women who were not in stable relationships, and did so with no knowledge of the principles of infectious disease and the contagion-limiting practices we know today. Thus, obstetricians at the time were commonly infected with syphilis by their patients. Semmelweis's behavioral changes are consistent with both medical diagnoses.

Other scholars say that his behaviors were the result of extreme frustration and emotional exhaustion from overwork and stress. Knowing he was clearly correct and yet still being rejected by the medical establishment with the resulting loss of prestige and damage to his livelihood, not to mention patients' lives, must have been agonizing. Whatever the explanation, in 1865 a colleague referred Semmelweis to a Viennese insane asylum. A second colleague lured him there under the pretense of showing him a new medical research institute. When Semmelweis realized what was happening, he fought against his involuntary admission and was severely beaten by several guards, secured in a straitjacket, and confined to a darkened cell. The injured physician lasted only two weeks and died of his wounds in August 1865 at the age of forty-seven. His autopsy stated the cause of death as septicemia (blood poisoning), the same cause of death produced by childbed fever.

Semmelweis was twenty-eight years old when he began his investigations into the cause and prevention of childbed fever and was thirty-one when he was discharged from Vienna General Hospital. He spent the remaining sixteen years of his life unsuccessfully trying to persuade the biomedical establishment to adopt the hospital infection control practices that he had developed.

When Semmelweis was buried, only a few people attended the service. Although it was standard practice to deliver a commemorative address on the death of a medical society member, this was not done for Semmelweis.

Semmelweis's successor at Szent Rókus Hospital in Budapest, János Diescher, did not believe in or follow Semmelweis's hand-washing procedure. Mortality rates in the maternity clinic quickly increased sixfold. During the following several decades, no one at Szent Rókus Hospital seemed to notice what was going on. There were no inquiries and no protests.

Fortunately, the germ theory was finally accepted several decades later, based in part on Semmelweis's own observations along with the further work of the English physicians John Snow and Joseph Lister, the French scientist Louis Pasteur, and the German physician scientist Robert Koch.

In recent years Semmelweis's rejection by the medical establishment has been investigated using approaches developed by the discipline of cognitive psychology. Scholars studying Semmelweis's career found that his difficulty in gaining acceptance for a breakthrough idea was not uncommon. These scholars named the phenomenon the "Semmelweis reflex," defining it as the "human behavioral tendency to stick to preexisting beliefs and to reject fresh ideas that contradict them, despite adequate evidence." It sounds a lot like Planck's principle.

There appear to be two psychological elements underlying the "Semmelweis reflex": belief perseverance (also called conceptual conservatism) and illusory correlation. *Belief perseverance* is defined as maintaining a belief in an idea despite the availability of new information that clearly refutes it. Maddeningly, studies of belief perseverance have shown that the belief in incorrect ideas is often strengthened when others attempt to provide evidence to disprove them.

As an example, in a recent study of vaccination hesitancy, subjects who worried about the side effects of seasonal flu vaccinations became less willing to be vaccinated after being told that the vaccine was completely safe. Paradoxically, the new knowledge made them distrust the vaccine even more, illogically supporting their prior mistaken belief.

The second psychological principle, *illusionary correlation*, is the phenomenon of perceiving a relationship between variables even when no such relationship exists. In Semmelweis's case, it was argued that doctors are knowledgeable and respected, and knowledgeable and respected physicians could not possibly spread infection. Of course, there is absolutely no correlation between professional stature and sterility, infection, or sepsis.

In the final analysis, Semmelweis's career made two important contributions to human knowledge. The work he performed during his lifetime contributed to the establishment of the germ theory of disease, which has saved countless lives. His work in addition provided a model for hospital infection control, lifesaving practices that are being followed every day in hospitals around the world. And years after his death, studies of how Semmelweis's ideas were rejected by the medical establishment during his lifetime have led to a better understanding of the cognitive psychology of why people reject correct ideas, with applications ranging from efforts to increase vaccination in support of public health to implementing measures that are needed to fight global warming.

Semmelweis differs from the men and women described in other chapters in that he was a clinician and not a scientist. The best medical practices are based on established scientific principles and are evidence-based. And although he was not a scientist, Semmelweis proceeded like one. He studied a number of variables to determine which, if any, were associated with childbed fever. He then developed a hypothesis and modified hospital procedures to show whether the variable under investigation was causal. And despite the fact that he was a clinician, he suffered the same fate as scientists who develop revolutionary new ideas: he was disbelieved, disapproved of, insulted, and finally shunned, despite having developed what we now know to be a major life-saving medical practice.

6 | Peyton Rous
Discoverer of the first cancer-causing gene

Cancer

It is said that cancer is as old as the human race, but paleontologists have found evidence of tumors in animals that lived prior to the appearance of humans on earth. Cancer is thus older than the human race. The earliest known written reference to cancer appeared in the Edwin Smith papyrus. Written in approximately 3000 BC, it described breast cancer, which we today know to be the most common human cancer.

During the succeeding several thousand years cancer was treated with primitive surgery and a variety of nonsurgical treatments ranging from herbal remedies (tea, fruit juices, figs, boiled cabbage) to toxic metals (iron, copper, sulfur, mercury, arsenic paste). None of this did much good. The main problem in finding an effective cure was that no one knew what caused cancer. Cancerous growths reminded the ancient Greek philosopher Hippocrates of a moving crab. Today we know that cancer of course has nothing to do with crabs, but the name stuck. We now use the Latin word for crab, *cancer*, as the name for the disease.

Hippocrates believed that diseases were caused by imbalances of what he called the four basic humors, fluids of the human body: blood, phlegm, yellow bile, and black bile. Different imbalances produced different diseases, cancer being no exception to this rule. The famous second-century Greek physician and philosopher Galen supported and promoted Hippocrates's theory. Despite the fact that it was neither true nor helpful in the treatment and management of disease, the humor theory became the most generally accepted explanation for disease until the mid-nineteenth century, when the germ theory of disease was developed. The germ theory explained that many diseases previously thought to be caused by humoral imbalances were actually caused by infection with microbes.

The late nineteenth-century German physician-scientist Robert Koch proved that many important diseases, including anthrax, tuberculosis, and cholera, were caused by microbial pathogens. He also established a set of criteria, now called Koch's postulates, by which physicians could prove whether a disease was caused by one. This development was critical, because knowing that a disease was caused by an infectious microbial pathogen made it possible to take measures to reduce the spread of the disease and later to develop highly effective treatments for many such diseases.

Scientists quickly embraced Koch's ideas and started to study a broad variety of diseases to determine whether or not they were infectious. One of these was a student of Robert Koch, a Danish scientist named Johannes Fibiger. Fibiger observed parasitic nematode worms in human stomach tumors and hypothesized that the worms had caused the cancer. He developed a rat model of stomach cancer to test his idea and published his results in 1913.

Fibiger fed rats a diet containing cockroaches that had been infested with nematode worms. Fibiger said that these rats later developed cancers of the stomach and esophagus, and he concluded that cancer is an infectious disease. Initially, his work was widely accepted and highly praised as providing a crucial insight into the cause of cancer. Fibiger was awarded the Nobel Prize for his discovery in 1926, a very rapid scientific validation for his findings. But soon researchers began reporting that they had difficulties in repeating Fibiger's observations. It turned out that the changes Fibiger had observed in the stomach and esophagus of his rats were in fact not cancers at all. Rather the changes he observed in the digestive tract lining were an inflammatory reaction to the presence of the nematodes. The worms did not cause cancer.

Why did the Nobel Committee make a unique error in awarding Fibiger a Nobel Prize for a discovery that soon turned out to be totally wrong? Perhaps they were influenced by the scientific fashion and style of the era. Many scientists were working on the idea that cancer might be an infectious disease and were publishing results that appeared to be consistent with Fibiger's theory. One of these was Francis Peyton Rous, but Rous was far, far more careful than Fibiger in interpreting his data and reaching conclusions.

Personal Life and Education

Francis Peyton Rous's mother was from Texas, and his father was from Maryland. After their marriage the couple settled in Baltimore. Rous's father died young, leaving his mother with three small children to care for and little money with which to do so. Rous's mother decided to stay on in Baltimore rather than move back to her family in Texas, feeling that her children would have better educational opportunities in Baltimore.

Peyton Rous, the eldest child, was born in 1879, received his college degree from Johns Hopkins University in 1900, and stayed on after graduation to earn a medical degree. But an accident during his second year of medical study almost killed him and could have deprived the medical scientific community of one of their greatest practitioners. Rous scraped the skin of one of his fingers on a bone from a tuberculosis patient while performing an autopsy. Soon a corpse tubercle formed on his finger. Rous had been infected with tuberculosis, a frightening diagnosis in 1902, a half century prior to the discovery of antibiotic treatments for the disease. In the early twentieth century 70 percent of the patients who were diagnosed with tuberculosis died. Rous's doctors did what little they could. The infection soon moved to the lymph nodes in his armpit, and these lymph nodes were surgically removed, but at the time for patients like Rous there were no further treatments.

In the absence of a curative treatment, tuberculosis patients were sent to sanitariums, places where patients with tuberculosis were isolated to prevent the spread of the infection and where they were treated with fresh air and sunlight. Basically, it was a "hoping for the best" approach. In Rous's case he was sent "to go away and try to get well" with his mother's family in Texas, though obviously this exposed them to tuberculosis in the process.

An uncle helped Rous get ranching jobs. Rous and a friend delivered two covered wagons full of hardware to Spur Ranch, an isolated community about 125 miles from the nearest railway station. Once they reached the Spur, Rous took a physically demanding job on a huge cattle ranch, riding horseback on cattle roundups and living and sleeping outdoors for days on end. The rigorous outdoor life provided a miracle cure. After one year Rous was able to return to Baltimore to complete his medical education.

Peyton Rous married Marion Eckford de Kay in 1915. They had three children and remained together until his death in 1970. One of their daughters, Marion, became a children's book editor and married Alan Hodgkin, a professor at Cambridge University who received the Nobel Prize in Physiology or Medicine in 1963 for his work on the electrical properties of nerve cells.

Research Studies

After finishing medical school in 1905, Rous decided to pursue a career in medical research rather than clinical medicine. He took a position as an instructor in pathology at the University of Michigan where the head of the department, Professor Alfred Warthin, strongly supported and encouraged his scientific ambitions. Warthin taught summer school in Rous's place and used the summer teaching earnings to provide Rous with a stipend to learn German and spend the summer of 1907 in Dresden, where he studied the latest advanced techniques in morbid anatomy. At this time, morbid anatomy was cutting-edge medical science. Cadavers were examined to identify anatomical changes that had resulted from their disease in the hope of discovering the underlying mechanisms by which the disease had killed the patient.

After Rous returned from Dresden, Professor Warthin encouraged him to continue his scientific training at the Rockefeller Institute for Medical Research (now called the Rockefeller University). There Rous trained under Simon Flexner, whose laboratory was studying cancer. Flexner later became interested in the viral disease poliomyelitis and asked Rous to take over his cancer work.

Rous's research career thus developed at the nexus of cancer and viral research. One day in 1909, a poultry farmer from nearby

Long Island contacted Rous. One of his barred Plymouth Rock hens had developed a lump protruding from its right breast. The farmer was worried that this bird might sicken other birds in his flock.

Rous biopsied the lump and found that it was a sarcoma, a connective tissue tumor. To address the farmer's worry that the disease might spread to other birds, Rous cut off a piece of the tumor, ground it up, and passed it through a filter to remove all cells. He chose a filter that would allow nothing larger than a virus to pass through it. He then injected the liquid into other chickens. Within a few weeks they too developed tumors. The farmer's worry was justified: the hen was suffering from a viral infectious disease.

Rous published his results in 1911, results that showed he had discovered a chicken sarcoma that appeared to be caused by a virus, later named after him, Rous sarcoma virus or RSV. But after publishing this result, Rous proceeded cautiously. Prior to Fibiger's publications, there was little to no support in the scientific literature for the idea that cancer was an infectious disease. Clinically, cancers appear to arise spontaneously. When Rous told colleagues he had identified a solid tumor that was produced by a virus, a skeptical scientist told him, "This can't be cancer, because you know its cause." Rous wanted to make certain he had not made some sort of error. He also worried that, even if what he found was real and not due to a mistake, it was potentially a scientific oddity that had nothing to do with cancer in general.

Rous tried to reproduce his finding with mouse cancers but was unsuccessful. A few years later, in 1915, he stopped working with tumors and did not return to cancer biology for almost twenty years, when a close friend and colleague, Richard Shope, asked Rous to collaborate with him on studies of a virus he had discovered. This virus, now called Shope papilloma virus, forms giant warts called keratinous carcinomas on the skin of wild rabbits.

These warts are benign tumors, but periodically they become cancerous, so, working with Shope, Rous had now found a second virus that caused cancer, this one in a mammal. Soon many other examples of tumor-inducing viruses were found in rabbits, mice, cats, and nonhuman primates. The first human oncogenic virus, Epstein-Barr virus, was described in 1964. With the isolation of the Shope papilloma virus along with the Rous sarcoma virus, it was now clear that, although unusual, it was possible for a virus to cause cancer. The question was, did these viral cancers have anything to teach us about the much, much more common spontaneous cancers?

Mechanism of Viral Cancer

Though Rous had discovered the cancer-producing RSV virus, he did not have the experimental tools to study it in depth and explain how it worked. In the late 1950s a group of scientists developed a new cell-based method to study RSV. As a result, laboratories that wanted to study the virus now no longer had to maintain a flock of chickens to conduct their experiments. This encouraged many new scientists to work on the virus and vastly accelerated the rate at which RSV research could be performed.

The new generation of RSV scientists soon identified a virus that was almost identical to RSV but did not cause cancer. There was therefore something very special about RSV that enabled it to cause cancer. Then a second virus almost identical to RSV was identified. This virus was unable to replicate and cause an infection, yet it still caused cancer. Thus the virus itself did not cause cancer, and neither did the process of viral infection and viral growth cause cancer. RSV appeared to carry a special kind of gene, a gene that was responsible for producing the sarcomas.

The gene was identified in 1970 and named *src*, short for the word *sarcoma*, the kind of cancer it produced. The next questions were: Where did the *src* gene come from and how did it work? In 1976 a group of American scientists, Michael Bishop, Harold Varmus, and Peter Vogt, made the surprising discovery that *src* was basically just an ordinary chicken gene. In fact, they found that *src* genes were present in a wide variety of birds including ducks, turkeys, quail, and even an emu from the San Francisco Zoo. The virus did not cause cancer. It was just a vehicle that transported a hitchhiking chicken gene, which then produced the cancer.

Since the chicken gene was the real culprit in producing cancer, researchers now pivoted to study the chicken gene to figure out how it worked. For a start, no one had any idea why the *src* gene was present in bird genomes in the first place and what role it had in the health and growth of normal cells.

The first major finding along these lines was that the *src* gene from the chicken was not identical to the *src* gene from the virus. While the *src* gene present in the RSV virus produced cancer, the *src* gene from chickens did not. It turned out that the *src* gene in RSV was a mutant version of the *src* gene present in the chicken, and the slight mutational change in the viral *src* gene brought about its cancer-causing properties.

Later research showed that there were many such genes, normal genes that became cancer-producing in a mutant form. Scientists called these genes oncogenes, genes that caused cancer. And the non-cancer-causing versions of these genes, normally present in chickens and other animals including humans, were called pro-to-oncogenes, genes that could change into oncogenes. But why did birds and animals carry these proto-oncogenes and what was their role in the life and growth of the birds and animals that carried them?

Oncogenes

To date more than seventy human oncogenes have been identified. Each gene is unique, but in studying them as a group it quickly became clear that they all were involved in the same process: the control of cell growth. In a multicellular organism, each type of cell must be present in the proper amount and in the right proportion to other types of cells. You cannot have too many or too few heart cells, or skin cells, or blood cells, etc. To achieve this, cell growth is tightly controlled by a system of highly regulated gene products.

Oncogenes encode proteins that control cell growth through a variety of mechanisms: growth hormones, cellular growth factors that selectively stimulate the growth of a specific type of cell, the receptors for these cellular growth factors, signal transduction factors that bring growth factor signals into the cells (*src* is of this type), and transcription factors that control how cells behave in response to growth factors. The activity of proto-oncogene forms is tightly controlled. But mutant oncogene forms escape control and become extraordinarily active, driving cells into rapid cancerous growth.

The *src* gene is the founding member of the oncogene gene family. Like other oncogenes, *src* affects cell proliferation and survival, adhesion to other cells and cell motility—exactly those properties that underlie the differences between the biology of normal cells and cancer cells. Normal cells grow but only under strict constraints. Cancer cells proliferate wildly and resist signals that damp down their growth and survival. Normal cells are restricted to their proper locations within the body. Cancer cells have altered adhesion properties that allow them to move to other parts of the body and create metastases when they arrive at new locations.

The *src* gene encodes a *protein kinase*, an enzyme that transfers phosphate groups to other proteins. The *src* protein is part of a tightly controlled signaling cascade, a process that transfers phosphate groups to a series of proteins like a bucket brigade, with each protein in the cascade transferring phosphates and thus gene growth information to the next protein in line until the target is reached, which then alters the way the cell behaves. This changed behavior results in cell proliferation and alterations to cellular properties that make the *src* gene–expressing cancer cell invasive.

Unlike the normal *src* protein, the mutated *src* protein becomes an unregulated accelerator, like a gas pedal stuck to the floor. It constantly pushes the cell to grow and divide. Scientists next asked how and why genes like *src* change into the oncogene form. Answering this question brought together whole lines of previously seemingly unrelated areas of cancer research.

Everything Comes Together—Carcinogens Are Mutagens
Domestic chimneys first appeared in Europe during the twelfth century and became common by the sixteenth to seventeenth centuries. Hot gases and smoke exhausting through a chimney produce a layer of creosote on the chimney walls, which can catch fire, often setting the entire building ablaze. In 1582 the Tudor royal family in England instituted an ordinance to prevent these

fires by requiring that chimneys be swept four times per year. They imposed a fine of 3 shillings and 4 pence for noncompliance. This ordinance drove the establishment and growth of the English chimney sweeping trade.

English chimneys were designed with small-diameter flues to create a better draught and followed convoluted routes as flues from multiple rooms were joined to discharge from a single chimney. Chimney sweeping work was therefore carried out by small young boys who climbed up the narrow chimneys, dislodging soot and creosote as they went. The boys spent all their waking hours covered with soot and, considering the hygienic practices of the day, probably their sleeping hours as well.

In 1775 an English surgeon named Percival Pott noticed that cancer of the scrotum often occurred in chimney sweeps in London, but rarely, if ever, in anyone else. The correlation between a rare disease and a relatively rare occupation suggested to Pott that there might be a causal relationship. Pott proposed that the cancer-causing agent was chimney soot that remained lodged in the furrowed folds of scrotal skin.

Pott's report was the first published account of an occupational cancer: a cancer caused by an agent in the workplace. A second example appeared one hundred years later from the work of the German scientist Richard von Volkmann. In the late 1800s von Volkmann studied the correlation between the exposure of workers in the German coal distillation industry to coal tars and the development of skin cancers. He proposed that coal tar caused skin cancer and encouraged investigators to identify the ingredient in coal tar that produced cancer.

In 1933, starting with two metric tons of coal pitch, James D. Cook at the Research Institute of the Cancer Hospital of London isolated seven grams of the carcinogen, a nearly one-million-fold enrichment from the starting material. The carcinogenic principle was the chemical benzo[a]pyrene. With the carcinogenic agent in coal tar now identified, the next question was to determine the mechanism by which benzo[a]pyrene produced cancer.

The clue came from a third report of occupational cancer: X-rays. The German physicist Wilhelm Roentgen discovered X-rays in 1895, and they quickly became an important diagnostic tool in medicine. Initially X-rays were thought to be safe and harmless and were liberally and casually used. For example, to test whether their equipment was working X-ray technicians routinely placed their hands in the X-ray beam to see whether it made their skin redden.

As early as 1902, physicians began to notice an increase in skin cancer among radiology technicians. This suggested the possibility that X-rays might be carcinogenic. This hypothesis was tested in 1908 by Pierre Edouard Jean Clunet, who showed that X-rays produced invasive cancers in experimental animals. Again, the question remained as to how X-rays produced cancer.

At this time X-rays were being studied in the totally unrelated field of genetics. Led by Nobel laureate Hermann Muller, geneticists showed that X-rays were mutagenic. This raised the questions as to whether X-rays produced cancer because they were mutagenic or whether they had two properties: one to produce cancer and another to produce mutants? Occam's razor—which states that a simple explanation is preferable to one that is more complex—argued that the simpler explanation, that cancer is caused by mutations, was probably the correct one.

Supporting this idea, many chemical carcinogens were also found to be mutagens. But a large number of chemical carcinogens were not mutagens, arguing for a casual association and not a causal one. One prominent example was benzo[a]pyrene from coal tar, which was undeniably a potent carcinogen and which just as clearly was not a mutagen. American biochemists James and Elizabeth Miller proposed a theory to reconcile the conflicting results. Enzymes in the liver chemically modify foreign toxic chemicals to inactivate them. But these two biochemists proposed that in this case perhaps the opposite was true, that liver enzymes could turn a harmless substance into a toxic one.

American biochemist Bruce Ames developed a system through which this idea could be very quickly and broadly tested. Ames employed a simple microbial system to determine the mutagenicity of chemicals. He paired this system with a protocol in which chemicals were preincubated with liver enzymes to produce the metabolites that are formed in animals. Benzo[a]pyrene and many other carcinogens that had previously been shown to be non-mutagenic became powerful mutagens after exposure to the liver enzymes.

And a final missing piece of the puzzle was soon discovered, a piece that fit perfectly and confirmed the mutation theory. This evidence came from the study of tumor suppressor genes. Tumor suppressor genes block tumor cell growth. These genes were now found to act by repairing genetic damage produced by mutagens, by inhibiting the growth and division of cells carrying mutations, and/or by inducing cell death when DNA repair is not possible. Tumor suppressor genes act directly against DNA mutations, clearly linking DNA mutation and tumorigenesis.

One of the best studied tumor suppressor genes in the scientific literature is p53, a gene which encodes a protein that was codiscovered by three laboratories, including the laboratory of

Arnie Levine, one of the professors on my PhD thesis committee. Disabling mutations in p53 are present in more than 50 percent of human cancers. The p53 protein guards against cancer in multiple ways: activating enzymes that repair DNA mutations, arresting the growth of cells with damaged DNA, and inducing the death of cells with irreparable DNA damage.

The tumor suppressor gene that is best known by the general public is BRCA1, breast cancer type 1 susceptibility protein. BRCA1 is expressed in the cells of breast and ovary, where it helps to repair damaged DNA and kill cells if damaged DNA cannot be repaired. Women carrying a mutation in the BRCA1 gene are unable to repair this DNA damage and have an up to an 80 percent risk of developing breast cancer and a 55 percent risk of developing ovarian cancer. (The average risk of a woman in the United States developing breast cancer during her lifetime is about 13 percent.)

Actress Angelina Jolie suspected that her family might carry a defective BRCA1 gene. Her mother, actress Marcheline Bertrand, had breast cancer and died of ovarian cancer. Her grandmother died of ovarian cancer. When testing showed that Jolie carried a defective BRCA1 gene she chose at the age of thirty-seven to undergo a preventive double mastectomy. Her aunt, who shared her BRCA1 defect, died of breast cancer just three months after Jolie's operation. And two years later Jolie had her ovaries removed to protect herself against ovarian cancer.

Thus, through the study of tumor suppressor genes, the circle was closed. Cancer is caused by mutant oncogenes. Mutagens introduce mutations that turn proto-oncogenes into oncogenes. Tumor suppressor genes protect cells against cancer by repairing the DNA damage produced by carcinogens and by killing cells with irreparable mutation damage. Patients with defective tumor suppressor genes exhibit a greatly increased chance of developing cancer.

Therapeutic Advances from Oncogene Science

These oncogene discoveries have led to the development of a new group of cancer treatments called targeted therapies. Prior to the discovery of oncogenes, no one knew what caused cancer. All that was known was that cancer cells divided more rapidly than most normal cells. Traditional cancer chemotherapies were therefore developed to simply kill very rapidly dividing cells. That was all medical scientists had to go on.

Of course, while cancer cells are very susceptible to these traditional chemotherapy drugs, there are many normal, healthy cells in the body that also divide rapidly. Rapidly dividing normal cells include skin cells, cells of the gastrointestinal tract lining, and cells in the bone marrow that produce red and white blood cells. The terrible side effects of traditional chemotherapeutic drugs— side effects that limits their dosage and effectiveness—are caused by the killing of these normal rapidly dividing cells. Hair loss occurs when hair follicle cells are killed; nausea and other digestive tract abnormalities are produced when gastrointestinal tract cells are killed; and anemia and immunosuppression occur when hematopoietic stem cells in the bone marrow are killed.

Fortunately, cancer cells differ from normal cells in that their rapid growth depends on the action of activated oncogene proteins, proteins that are not activated in normal cells. Drugs that act to inhibit these oncogene products and/or the pathways they regulate are selective, better tolerated than traditional therapies, and can thus be dosed more aggressively than traditional cancer drugs. Oncogene-based targeted therapy drugs are also precisely matched to the oncogenes specifically driving the growth of the patient's tumor.

As an example, the growth of chronic myelogenous leukemia cancers depends upon an oncogene that encodes constitutively active tyrosine kinase enzymes in these cancer cells. Gleevec (imatinib) selectively inhibits these abnormally active signal transduction enzymes. Gleevec can thus be dosed very aggressively to treat these tumors, with much more manageable side effects and much greater efficacy.

A second example involves the treatment of solid tumors. The hormone vascular endothelial growth factor (VEGF) stimulates the growth of new blood vessels. Mutant oncogenes cause tumors to produce VEGF to induce the growth of blood vessels into the growing tumor cell mass. These new blood vessels bring the growing tumor cells needed oxygen and nutrients, without which they would be unable to grow beyond the tiny size of 1–2 millimeters. The formation of new blood vessels in and around these tumors is essential for the tumor's ability to grow exponentially. Avastin (bevacizumab) targets and inactivates VEGF. By so doing, the drug prevents solid tumors from becoming vascularized and blocks their growth. Approved on 2004, it is a highly effective treatment for VEGF-dependent colon, breast, and non–small cell lung cancers.

As a third example, breast cancer oncogenes cause the HER2/neu receptor to be overexpressed in approximately 30 percent of breast cancers. The overexpression of this receptor causes these tumors to grow rapidly. Herceptin (trastuzumab) blocks the HER2/neu receptor and can be dosed with far fewer side effects than conventional cytotoxic therapies. Herceptin greatly improves overall survival in both late-stage (metastatic) and early stage HER2-positive breast cancer, where it also reduces the risk of cancer returning after surgery

This is just a very short list of some of the best-known targeted therapy drugs. Thus far, almost one hundred conventional small molecule and two hundred monoclonal antibody targeted drugs have either been FDA approved or are in clinical trials for cancer treatment.

Cancer and Peyton Rous's Legacy

RSV is a freak of nature. Somehow an ordinary chicken virus picked up a potent oncogene and began ferrying it around. As a superb, inquisitive scientist, Peyton Rous saw something that he could not understand and set about to explain it. While Rous did not have the technical tools available to him that were needed to completely solve the problem, he proceeded as best he could, carefully collecting and cautiously interpreting his data, and thus establishing a solid foundation to enable the work of later scientists.

Once the required research tools were developed, later generations of scientists eventually figured everything out. For his discovery of tumor viruses, Rous was awarded the Nobel Prize in Physiology or Medicine in 1966, fifty-five years after his original findings. He received the award at the age of eighty-six, just four years prior to his death, and he remains the oldest recipient of the Nobel Prize in Physiology or Medicine. (The 2018 Nobel Prize in Physics was awarded to ninety-six-year-old Arthur Ashkin, and the 2019 Nobel Prize in Chemistry was awarded to ninety-seven-year-old John B. Goodenough.)

Rous's solid foundational research also spawned a series of subsequent Nobel awards. Based on Rous's pioneering work, the 1975 Nobel Prize in Physiology or Medicine was awarded to Renato Dulbecco, David Baltimore, and Howard Temin for their studies on the interaction between tumor viruses and the genetic material of the cell. Almost sixty-five years after Rous's discovery, this Nobel Prize cited the 1975 recipients for providing the conceptual foundation and technology needed to establish a relationship between viruses and human cancer. These three Nobelists showed that the critical element was the ability of the virus to interact with the DNA of the host.

The *src* gene was finally fully characterized and sequenced during the early 1980s, slightly more than ten years after Rous's death. And slightly more than seventy-five years after his initial experiments, the 1989 Nobel Prize in Physiology or Medicine was awarded to Michael Bishop and Harold Varmus for their discovery that *scr* was derived from a normal chicken gene, thus launching the new oncogene theory for the mechanism causing cancer.

And almost a hundred years after Rous's first RSV work, finally turning full circle, the Nobel Prize in Physiology or Medicine in 2008 was awarded for the discovery of human immunodeficiency virus (HIV) by Françoise Barré-Sinoussi and Luc Montagnier. HIV, the retrovirus that causes AIDS, is a member of a virus group whose founding member was none other than Rous sarcoma virus. The award was shared by Harald zur Hausen, who showed that human papilloma viruses cause cervical cancer.

Rous had to wait more than half a century for his achievement to be recognized. Although his work is now taught to undergraduate students as a foundational principle of cell biology, in the early twentieth century, it was beyond controversial. Rous lacked the technical tools to conclusively solve his problem, so instead he laid a solid scientific foundation for others to build upon. None of our current advances could have been made without him. The incorrect awarding of the Nobel Prize to Johannes Fibiger had cast a pall over Rous's work and set the whole field of cancer research back for many years. There were doubters and opponents even decades after the overall acceptance of the onco-gene hypothesis.

I wrote in the introduction to this book that in the 1980s I had proposed developing antioncogene protein therapies for the treatment of cancer. My employer hired the famous cancer researcher, Sir Henry Harris, as a consultant to assess my idea. Sir Henry concluded that my idea was scientifically unsupportable and destined to fail, and he stated that the scientists upon whose work I had based my proposal were about to retract their findings. Instead, some of the scientists whose work I had cited went on to win Nobel Prizes. Some of the best contemporary cancer treatments act against oncogene proteins. But Sir Henry remained a devoted oncogene opponent and skeptic until his dying day in 2014, thus providing a classic example of Planck's principle.

7 | Roger Revelle
Climatologist and policy advocate whose
findings were the first to be described by
the term "global warming"

Climate Change

Greek philosophers speculated as to whether the Earth's climate might have changed but without much evidence or theoretical basis. Serious scientific inquiry into the possibility of large-scale prehistoric climate change began only in the late eighteenth century.

Jean-Pierre Perraudin was a chamois hunter who hiked the hills and valleys of the Swiss canton of Valais in search of game. In 1815, as he hiked in the Val de Bagnes, he realized that it must have taken some enormous force to scatter the giant granite boulders he saw strewn all around the narrow valley floor. He also took note of a series of long stripes along the exposed rocks that lined the narrow valley. Perraudin proposed that ice from receding glaciers might have gouged out the stripes and also scattered the huge granite boulders down into the valley.

The Val de Bagnes cuts into the northern face of the southern Alps, running northward to the Rhone River Valley. Comparing early maps of Valais with their modern-day equivalents, one can clearly see that the Valais canton glaciers have dramatically retreated since Perraudin made his observations two hundred years ago. In Perraudin's time, the edge of the Corbassière glacier was high up in the mountains far above the Val de Bagnes, perhaps some ten kilometers from the valley floor. Today it has receded several kilometers higher up into the mountains from where it was in Perraudin's time.

Another Swiss glacier, the Lower Grindelwald Glacier, sits on the northern slope of the Swiss Bernese Alps, about 150 kilometers north and east from the Val de Bagnes. In the early nineteenth century, the Lower Grindelwald Glacier touched the outskirts of the town of Grindelwald. When I first hiked the Lower Grindelwald Glacier in August 1992, the trailhead was about a five-minute drive south from the village. A twenty-minute hike from the trailhead parking area took you to the glacier edge. When I returned in August 2009, a two-hour hike from the trailhead took me to a point where I could just see the glacier in the far distance. Maps show that the Lower Grindelwald Glacier has shrunk by about 25 percent in just the last fifteen years.

Perraudin's idea was initially met with general disbelief. But twenty years later, in 1837, Jean Louis Rodolphe Agassiz, a Swiss paleontologist at the University of Neuchâtel, studied the mountainous region of southern Switzerland where Perraudin had lived, and he became convinced of the merits of Perraudin's glacial theory. Agassiz proposed that the Earth had been subjected to a past ice age, that ancient glaciers had flowed outward from the Alps, that even larger glaciers had covered the plains and mountains of Europe, Asia, and North America, and that the entire Northern Hemisphere had been extensively covered with glaciers in a prolonged ice age. But Agassiz provided no explanation as to how or why this had happened.

At about this time the French mathematician and physicist, Jean-Baptiste Joseph Fourier, wrote that the Earth's atmosphere kept the planet warmer than it would have been in a vacuum. He reasoned that the atmosphere transmitted visible light waves efficiently to the earth's surface, which then absorbed the light and emitted infrared heat radiation in response. But since the atmosphere does not transmit infrared efficiently, this would cause an increase in surface temperatures. In 1838 the physicist Claude Pouillet first proposed (but without experimental evidence) that water vapor and carbon dioxide (CO_2) in the atmosphere might be responsible for trapping infrared light and warming the planet.

Evidence in support of this idea was soon obtained. In 1856 American scientist Eunice Newton Foote performed experiments showing that the warming effect of the sun was greater for air than a vacuum and greater for moist air than dry air, and that the greatest effect was for air containing high levels of CO_2 gas. She speculated that in the past Earth temperatures might have varied as a consequence of changes in atmospheric carbon dioxide levels.

In 1896 Swedish scientist Svante Arrhenius calculated the relationship between the level of atmospheric CO_2 and atmospheric temperature. He concluded that cutting atmospheric CO_2 in half would produce an ice age while the doubling of atmospheric CO_2 would produce a warming of 5–6 degrees Celsius. However, while it was clear that humans burning fossil fuels like coal would increase atmospheric CO_2, almost no one in the scientific community at the time thought that the puny effects of human activity could affect the vast climate cycles and, even if they could, major change seemed impossible except over the span of tens of thousands of years.

Roger Revelle

Roger Revelle was born in Seattle, Washington, in 1901 and grew up in Southern California. He studied geology at Pomona College and then earned a doctorate in oceanography from

University of California, Berkeley. As a Berkeley graduate student, Revelle performed his thesis research at the Scripps Institution of Oceanography in San Diego, studying deep-sea mud, and completed his doctoral thesis in 1936, "Marine Bottom Samples Collected in the Pacific Ocean by the Carnegie on Its Seventh Cruise."

The Scripps Institution of Oceanography was founded in 1903 by William Ritter, a Berkeley professor who wanted to establish a coastal research laboratory where he could collect specimens of marine life while leading summer classes. Ritter's idea of establishing a San Diego laboratory was supported by a group of local residents, who in 1903 formed the Marine Biological Association of San Diego, naming Ritter the scientific director. The association raised money for the construction of a small laboratory at La Jolla Cove.

But this fledgling laboratory was soon transformed into a major oceanographic research institution through the philanthropy of Ellen Browning Scripps and her brother, the newspaper magnate E. W. Scripps. The Scrippses purchased a 170-acre parcel on the La Jolla coast and funded the construction of the first laboratory, the iconic Scripps pier, and Ritter's residence at the laboratory. They bankrolled the laboratory's early operating expenses. In 1912, the Scripps Institution of Oceanography was named after the family whose financial support had made it a reality. In addition to what eventually became his long-term professional association with Scripps, while at Berkeley Revelle married Ellen Clark, a grandniece of E. W. Scripps and Ellen B. Scripps.

After completing his doctoral studies, Revelle spent a year at the Geophysical Institute in Bergen, Norway, before returning to Scripps. While working there during the 1930s, Revelle also served as an officer in the Naval Reserve. He went on active naval duty six months prior to the bombing of Pearl Harbor and stayed in the Navy for the next seven years. One of his Navy assignments was as the officer in charge of geophysical measurements during the 1946 atomic tests at Bikini Atoll. For years afterward Revelle maintained a high level of interest in the environmental effects of radiation and questioned the disposal of atomic wastes at sea.

After the war Revelle was instrumental in establishing the University of California San Diego (UCSD) as a companion institution to the Scripps Institution of Oceanography, working to have the new university sited on 1,100 acres of largely undeveloped public land just to the north of Scripps. He remained at Scripps and UCSD until 1961, when he became science advisor to Secretary of the Interior Stewart Udall. After his time at the Department of the Interior, Revelle was appointed the Richard Saltonstall Professor of Population Policy at Harvard University, where he taught for just over a decade. Revelle returned to UCSD in 1975 to become professor of science and public policy, and he taught courses in marine policy and population for the next fifteen years.

Revelle died in 1991 at the age of eighty-two. Just prior to his death, Revelle was awarded the National Medal of Science by President George H. W. Bush. He told a reporter, "I got it for being the grandfather of the greenhouse effect."

Climate Research

In the late 1950s Revelle established a collaboration with Hans Suess, an Austrian physicist who was an expert on carbon isotopes. Carbon exists in three forms: ^{12}C, ^{13}C, and ^{14}C. Of these three forms, ^{12}C and ^{13}C are stable while ^{14}C decays into a form of nitrogen. The ^{14}C isotope is continuously formed by the action of cosmic rays in the outer atmosphere. Thus, in the absence of other influences, the amount of ^{14}C in the atmosphere would exist in a steady state, depleted by decay but also replenished by the action of cosmic rays.

Suess showed that this steady state would be disturbed by the introduction into the atmosphere of ancient, sequestered carbon. Ancient carbon, the carbon in fossil fuels, is depleted of ^{14}C by decay but is not replenished by the action of cosmic radiation. Ancient carbon has less ^{14}C than contemporary atmospheric carbon. Thus when this ancient carbon is introduced into the atmosphere, it will disturb the steady state, leading to a decline of the proportion of ^{14}C. This is now called the Suess Effect. The decline in ^{14}C by decay in sequestered carbon is also the basis for the carbon dating of ancient materials of biological origin.

In a critical paper published in 1957, Revelle and Suess showed that reported increases in atmospheric CO_2 were due to the introduction of ^{14}C depleted carbon. The only reasonable explanation for these increases in atmospheric CO_2 levels was the burning of fossil fuels by humans.

Revelle's background in oceanography gave him a special perspective on climate change science. It had long been proposed that much of the CO_2 emissions produced by human activity would be absorbed by the ocean instead of entering the atmosphere. But Revelle used his knowledge of ocean chemistry to show that CO_2 can only enter the ocean by combining with water to produce three forms of carbonic acid. This results in a kind of back-pressure that limits how fast the carbon dioxide can enter the ocean surface, now called the "Revelle factor." As the ocean surface layer thus has a limited ability to absorb carbon dioxide, most CO_2

emissions end up in the atmosphere, with troubling potential effects on climate change. In addition, CO_2 that is absorbed by the ocean leads to the acidification of ocean water with potential serious negative consequences for many marine species.

After the publication of his 1957 paper on ocean CO_2, Revelle told journalists that "the Earth itself is a space ship," endangered by rising seas and desertification. Revelle wanted everyone to keep a better eye on the spaceship's air control system. Newspapers at the time reported that Revelle's research indicated that "a large-scale global warming, with radical climate changes, may result."

When Revelle started his career, anthropogenic (human-caused) global warming was a fringe idea. But unlike other scientists described in this book, Revelle was quickly able to gain the support of the scientific community for his ideas. Recent surveys on the level of scientific consensus among climate experts on the idea of human-caused global warming have consistently shown agreement to be at or above the 90 percent level, and a 2019 study reported the scientific consensus to be at 100 percent.

Not so, however, with nonscientists. While the scientific community is close to or at complete agreement on the issue of anthropogenic global warming, even the most optimistic current assessments, such as a recent Pew Research Center study, indicate that some 40 percent of American adults believe that climate change will not harm them personally during their lifetimes.

Revelle's long-term objective was not simply to establish a new improved climate theory. Rather, he wanted his new theory to be used to protect humanity and the planet Earth from the damage and destruction that would be produced by uncontrolled climate change. And he realized that in order to achieve this goal, he would have to switch his efforts from science to public policy.

Public Policy

Throughout his career Revelle took a global perspective on earth science investigations and advocated for the broad application of earth science in everyday life. While working for the Navy, he encouraged them to support basic research in addition to working on projects with direct application to issues of importance to the military. After the war, he promoted an international approach to the study of oceanography, including long-range expeditions involving collaborations with other countries. Together with scientists at the Woods Hole Oceanographic Institution in Massachusetts, Revelle worked on American contributions to the oceanographic program of the 1957–1958 International Geophysical Year. At the same time, he became the first president of the Scientific Committee on Oceanic Research, an international group of scientists devoted to advising on international projects.

Revelle realized that the collective efforts of all of humanity would be needed to address the dangers of global warming. His public policy efforts focused on the creation of the International Geophysical Year, and he also served as the founding chairman of the first Committee on Climate Change and the Ocean. Under Revelle's directorship, the Scripps Institute for Oceanography became the principal center for the Atmospheric Carbon Dioxide Program.

Continuing his public policy efforts, in 1964 Revelle founded the Center for Population Studies at Harvard University, spending the next decade as its director. Among his students at Harvard were Benazir Bhutto, who later became the prime minister of Pakistan, and Albert Gore, who later became the vice president of the United States under Bill Clinton. In the late 1970s Revelle returned to Southern California, concentrating his efforts in public policy and education, teaching systems theoretic process analysis seminars on energy and development, the carbon dioxide problem (global warming), and marine policy.

Legacy

Revelle differs from other scientists in this book in that his idea was reasonably quickly embraced by the scientific community, perhaps in no small part because it was easy to understand and was supported by good common sense. Revelle was fortunate during his lifetime to make a major impact on the field of climatology, taking a fringe idea—anthropogenic global warming—into the scientific mainstream. But it soon became clear that the major challenge was to get the rest of humanity to accept this idea and for the world to take the steps needed to halt or reverse the process. This turned out to be much more difficult than convincing climate scientists.

People who disagree with the scientific consensus call themselves "climate change skeptics." But scientists consider this label to be incorrect. A skeptic is someone who disagrees with the commonly held viewpoint but keeps an open mind, awaiting confirmation or disproval by new facts. Scientists pride themselves on being skeptics until unequivocal evidence is accumulated to prove a new idea. Thus, scientists feel that the correct term for such individuals is "climate change denialists," *denial* meaning the rejection of well-established facts in order to support a contrarian view.

Models of anthropogenic global climate change make dire predictions, the most extreme of which are a significant die-off of the human population and possibly the extinction of the human species. Why would anyone deny something with high-stakes existential significance? One answer could be money. The fossil fuel industry has enormous reserves of coal, oil, and natural gas. At current market prices, the industry holds about $600 trillion of coal, $90 trillion of oil, and $70 trillion of natural gas reserves. To put these head-spinning numbers into context, the entire total annual value of all goods and services produced in the United States is about $21 trillion. Thus, the total reserve holdings of the fossil fuel industry is about thirty-five years of total US output. The industry does not want to take the financial hit of writing this asset down should fossil fuel use be banned to prevent further global warming.

The groups opposed to accepting the science of anthropogenic global climate change include the fossil fuel industry and industries whose business models are fossil fuel dependent; their conservative lobbyists; certain think tanks including Freedom-Works, Americans for Prosperity (funded by brothers David and Charles Koch of Koch Industries), the Heritage Foundation, the Marshall Institute, the Cato Institute, the American Enterprise Institute; and the Republican Party.

Tactics against anthropogenic global climate change science include the support of bogus "experts" who present themselves as scientists but are in reality simply political advocates working in behalf of anti–climate change supporters. Another tactic is to paint global warming scientists as no more than a group of dishonest coconspirators with the goal of suppressing the so-called truth about the climate. Yet anyone who has any experience with science knows that scientists are lone wolves, working to show that their ideas are correct and their scientific competitors' theories are wrong. Scientists do not easily come to work together. Getting them to participate in a conspiracy is unimaginable.

Selectivity is another tactic, elevating rare, unusual anti–climate change publications and events as providing the true story on climate science. February 2015 was an unusually cold month in Washington, DC. There was snow on the ground. James Inhofe, a senator from Oklahoma, the chairman of the Senate's Environment and Public Works Committee, and a devoted climate change denier, brought a snowball into the Senate chamber as proof that global warming was a hoax. Inhofe promoted this rare Washington, DC, snowball as a clear refutation of decades of work by the scientific climatology community. But unusual events neither prove nor disprove scientific theories.

Deniers in many fields use selectivity, cherry-picking atypical papers that happen to agree with their viewpoint, as a major tactic. Andrew Wakefield published a paper on autism in 1998 in the highly regarded British journal *Lancet*. The paper described children developing autism following an injection of MMR (measles, mumps, and rubella) vaccine. But twelve years later the paper was retracted by the journal, an almost unheard-of development. The meticulous review process of journals like the *Lancet* almost always prevents papers with incorrect or dubious findings from being published in the first place. According to the journal, the paper was retracted because "several elements are incorrect, contrary to the findings of an earlier investigation." Wakefield was then stripped of his medical license by the British General Medical Council, who found that he had been dishonest in his research, had acted against his patients' best interests, and had mistreated developmentally delayed children. Despite this, more than ten years following the retraction and Wakefield's loss of his medical license, the Wakefield paper is still commonly cited as definitive evidence that vaccines cause autism.

Another tactic is the imposition of impossible demands on scientists who support the theory of anthropogenic global climate change. A fundamental basic principle of science is that one cannot prove something to be impossible. Denialists propose that the currently observed climate change could be caused by volcanism. Since it is impossible for scientists to disprove with certainty that volcanism is causing climate change, opponents use this as an example showing that causes other than human activity, like volcanism, are just as likely to explain climate change.

A fifth tactic is logical fallacies. Opponents say that there have been major climate changes in the past in the absence of human activity, and therefore human activity does not cause climate change. But evidence that in the past someone crashed their car while driving sober does not prove that a current car wreck is unrelated to the driver's heavy drinking just prior to the crash.

Republican opposition to any policies or legislation to address climate change has been uniform and unrelenting. During the 2016 United States election cycle, *every* Republican presidential candidate questioned or denied climate change and opposed US government steps to address it. Ronald Reagan was arguably the first Republican president to make denial of global warming a major part of his political platform. Reagan promised to cut spending on environmental research, particularly climate-related research, and to stop funding for CO_2 monitoring. Reagan appointed James B. Edwards, a global warming denier, as secretary of energy. When the Environmental Protection Agency (EPA) reported in 1983 that global warming was "not a theoretical problem but a threat whose effects will be felt within a few years," with potentially "catastrophic" consequences, the Reagan administration called the report "alarmist" and disputed the EPA position.

The Reagan administration position was opposed by Congressman Al Gore, among others. Gore's position on the issue was formed by a class he took at Harvard, taught by none other than Professor Roger Revelle. Gore and other like-minded members of Congress arranged hearings on the issue and had credible climate scientists, including Revelle, testify before Congress. In 1988 James Hansen, a professor directing the Program on Climate Science, Awareness and Solutions of the Earth Institute at Columbia University, testified at a congressional hearing, stating with high confidence that long-term warming was underway, with severe warming likely within the next fifty years, and warning of likely storms and floods. President Reagan's rejectionist climate policies were continued by the administration of President George H. W. Bush.

In 1992 Democrat Bill Clinton was elected president and climate protectionist Al Gore vice president. The many climate-friendly actions taken by the Clinton administration included the Climate Change Action Plan to reduce greenhouse gas emissions to 1990 levels by 2000; the BTU (British thermal unit) tax on producers of gasoline, oil, and other fuels; the joint implementation in the 1995 Conference of the Parties (COP-1) calling for developed countries to lead the implementation of national mitigation policies; the signing of the Kyoto Protocol on behalf of the United States, pledging the country to a nonbinding 7 percent reduction of greenhouse gas emissions; the FY00 proposal, which allotted funding for a new set of environmental policies including a new Clean Air Partnership Fund with tax incentives, investments, and funding for environmental research of both natural and manmade changes to the climate; the Clean Air Partnership Fund to finance state and local government efforts for greenhouse gas emission reductions in cooperation with EPA; and the Climate Change Technology Initiative, which provided $4 billion in tax incentives for building energy efficient homes, installing solar energy systems, manufacturing electric and hybrid vehicles, and supporting clean energy production by the power industry and transportation sectors.

Vice President Al Gore ran for the presidency in 2001 against George W. Bush. Gore won the popular vote by approximately 500,000 votes. But winning the electoral college turned on the vote in the state of Florida, where George W. Bush defeated Gore by 537 votes. Thus, Bush became the next president of the United States. Bush returned to Republican climate policies, reneging on the Kyoto Protocol and, according to congressional testimony, pressuring American scientists on behalf of the fossil fuel lobby to suppress discussion of global warming and to alter their publications by eliminating terms such as "climate change" and "global warming." Some government scientists resigned their positions rather than submit to this censorship. Climate science currently predicts that much of the state of Florida will lie below sea level if nothing is done to halt global warming. This appears to be what the citizens of Florida voted for.

After his defeat, Al Gore, former Roger Revelle student, focused his efforts on environmentalism. In 2004 he colaunched Generation Investment Management to create environment-friendly portfolios for investors. Gore and the Alliance for Climate Protection created the "We Can Solve It" organization, a web-based program, to spread awareness of climate change to help end global warming. Gore starred in the 2006 documentary film *An Inconvenient Truth*, which warned about the consequences of global warming. The film is the fourth-highest-grossing documentary in US history. In 2007 Gore was awarded the Nobel Peace Prize together with the Intergovernmental Panel on Climate Change "for their efforts to build up and disseminate greater knowledge about man-made climate change, and to lay the foundations for the measures that are needed to counteract such change."

Democrat Barack Obama was elected president following the end of George W. Bush's second term. In what has become the standard seesaw behavior of American climate policy, President Obama pursued a large number of climate-friendly initiatives to combat climate change. Shortly after being elected, President Obama addressed the international community on climate change, saying "we risk consigning future generations to an irreversible catastrophe . . . our prosperity, our health, and our safety are in jeopardy, and the time we have to reverse this tide is running out." In April 2016 the United States joined the international Paris Agreement to fight climate change.

Major domestic Obama climate change programs included the establishment of a new office, the White House Office of Energy and Climate Change Policy; the institution of the 2015 Clean Power Plan to reduce carbon dioxide emissions from power plants; the development of plans for a 40–45 percent reduction in methane emissions by 2025 (methane emissions are responsible for up to one-third of near-term global heating); and negotiating an agreement with thirteen large automakers to increase fuel economy to 54.5 miles per gallon for cars and light-duty trucks by model year 2025.

After his election, President Donald Trump reversed or cancelled virtually all of the Obama administration environmental plans and policies. He withdrew from the Paris Agreement, saying that its terms would damage the US economy and its world competitiveness. He rolled back the Obama fuel economy standards, saying that Americans preferred to drive large cars and should not be compelled by the government to purchase smaller, less safe vehicles. He reversed President Obama's clean power plan, advocating coal-based electricity generation to support American energy independence and good-paying American jobs. He rolled back the Obama administration's methane emission rules as well.

In August 2022, under President Joe Biden, thirty years after Roger Revelle's death, the United States Congress passed the Inflation Reduction Act by a razor-thin margin, with every Republican in both chambers voting against it. The bill provides $370 billion to begin to move the country away from fossil fuels and toward solar, wind, and other renewable energy sources. It was the nation's first major climate law.

According to Democrats, the bill will lower greenhouse gas emissions by 40 percent in the next ten years relative to 2005 levels. An analysis by the REPEAT Project at Princeton University, an academic group that studies climate data and climate legislation, estimates that a 50 percent reduction will be possible if the measures put in place by the Inflation Reduction Act policies are combined with supportive executive action.

Climate science says that a sustained and unified approach is needed to battle climate change. Our current scattershot method, with dramatic reversals every four to eight years as new presidents from different parties are elected, will likely lead to severe climatic changes with extreme negative consequences for humanity. Revelle understood this and for that reason focused his later efforts on public policy.

Revelle was successful in getting the academic community to embrace his theories on climate change. But humanity at large has yet to coalesce to accept his ideas. Since Revelle's death in 1991, average global temperatures have increased by slightly more than 0.5°C (about 1°F), ten of the warmest years ever recorded have occurred, and sea level has increased by about 90 millimeters (3.5 inches). Climate science predicts that if nothing is done, these effects will accelerate rapidly.

Climatologists are in agreement. While the passage of the Inflation Reduction Act is widely regarded as an excellent start, it remains to be seen whether Revelle's ideas will become generally accepted by politicians and policy makers, not to mention the population at large. A broad worldwide consensus is needed to turn his scientifically accepted ideas into actions that will protect the well-being of humanity and Planet Earth.

8 | Rachel Carson
Catalyst for the environmental movement

Rachel Carson was born in 1907 and raised on a family farm on the Allegheny River in western Pennsylvania, just east of Pittsburgh. Her lifelong passions were the natural world and writing. An avid reader, she attended Chatham University in Pittsburgh after high school, majoring in biology and writing for the school's newspaper and literary magazine. She had hoped to transfer to the more prestigious Johns Hopkins University but was unable to do so because of her family's precarious financial position.

After graduation she was awarded a scholarship to attend the Johns Hopkins graduate school biology program and earned a master's degree there in 1932. Carson wanted to continue at Johns Hopkins to earn a doctorate but was forced to leave and take a job to help support her family during the Great Depression. In 1935 her father died suddenly, leaving Carson to care for her aging mother. Carson took a temporary position with the US Bureau of Fisheries, writing radio copy for a series of weekly educational broadcasts titled "Romance Under the Waters." Jobs were hard to come by during the Great Depression, and this position had the advantage of combining her two passions: biology and writing.

Carson wrote about aquatic life, marine biology, and the work of the US Bureau of Fisheries. She also began writing articles about marine life in the Chesapeake Bay for local newspapers and magazines, especially the *Baltimore Sun*. In 1937 her older sister died, producing further economic stress. Carson now became the sole breadwinner for her mother and two nieces.

In July 1937 the *Atlantic Monthly* accepted an essay she had titled "The World of Waters," which was based on material Carson had originally written for the US Bureau of Fisheries as an eleven-page introduction to a government fisheries brochure. It was an account of an ocean floor journey and appeared in the magazine as "Undersea." The publishing house Simon & Schuster was so impressed by "Undersea" that they contacted Carson to propose that she expand it into a book, which appeared in 1941 with the title *Under the Sea Wind: A Naturalist's Picture of Ocean Life*.

Under the Sea Wind described the life of organisms living in the sea and on the Atlantic coast over the span of a year. The first section tells the story of a female sanderling, which Carson named Silverbar. The second section does the same for a mackerel she named Scomber, and the third section follows the life of an eel. The book viewed ocean life from a broad ecological perspective, showing how each of these creatures existed within the broader web of life, and warned how humans are increasingly becoming separated from nature.

After the World War II, Carson planned to leave the US Bureau of Fisheries, now renamed the United States Fish and Wildlife Service, to pursue a full-time literary career. In 1948 Carson began working on a second book, which was published in 1951 as *The Sea Around Us* by Oxford University Press. *The Sea Around Us*, which provided a loving description of the oceans and all the remarkable and mysterious creatures that live therein, was a great success. Chapters appeared in *Science Digest* and the *Yale*

Review. The *Yale Review* chapter, titled "The Birth of an Island," won the American Association for the Advancement of Science's George Westinghouse Science Writing Prize. Nine chapters were serialized in *The New Yorker* in 1951.

The Sea Around Us remained on the *New York Times* bestseller list for eighty-six weeks, selling well over a million copies, and won the 1952 National Book Award for Nonfiction and the John Burroughs Medal. A film version won the 1953 Oscar for Best Documentary. Afterward Carson was awarded two honorary doctorate degrees. The success of *The Sea Around Us* led to the republication of *Under the Sea Wind*, which soon also became a bestseller too. In 1952, with the income from two bestselling books, Carson was finally able to give up her job with the United States Fish and Wildlife Service to concentrate on writing full time.

In early 1953 Carson began library and field research on the ecology and organisms of the Atlantic shore for the third volume in her sea trilogy, *The Edge of the Sea*, a book that would describe life in coastal ecosystems, focusing on the Eastern Seaboard. The book appeared in *The New Yorker* in two condensed installments and was released by Houghton Mifflin on October 26, 1955. *The Edge of the Sea* became Carson's third critically acclaimed book.

After the publication of *The Edge of the Sea* Carson began considering subjects for a fourth book. Her interests turned to conservation and environmental threats. She became alarmed by federal proposals from the US Department of Agriculture for widespread pesticide spraying, which they had recommended to eradicate fire ants. The dangers of pesticide overuse became Carson's principal literary focus for the rest of her life.

This concern became the subject of her masterwork and best-known book, *Silent Spring.* Published by Houghton Mifflin in

September 1962, it described the harmful effects of pesticides on the environment and is generally credited as having launched the environmental movement.

Agricultural Pests and Pesticides

There are three types of agricultural pests: weeds, fungi that cause plant disease (phytopathogenic fungi), and insects.

Weeds

Weeds compete with crop plants for soil nutrients, space, and light. Historically, weeds were dealt with by picking—mechanical removal. With time, mechanical removal was improved with the development of tools like the garden hoe. Both the crop and the weeds are plants. While hoeing can be successfully carried out with minimal manual dexterity, other methods of weed removal face the conundrum of selectively killing one plant without injuring another.

The solution and the first herbicide, weedkiller, was discovered simultaneously in England and the United Sates during World War II by four groups working independently under wartime secrecy. This herbicide was 2,4-Dichlorophenoxyacetic acid, commonly called 2,4-D. Indole-3-acetic acid (IAA) is the most common and important naturally occurring plant hormone of the auxin hormone class. 2,4-D is a chemical analog of IAA, which is metabolically and environmentally much more stable than IAA itself. When applied to plants, 2,4-D disrupts hormonal signaling, killing the plant.

What makes 2,4-D a useful herbicide is that higher plants belong to two different families: dicotyledon plants, also known as dicots or broad leaf plants, and monocotyledon plants, commonly referred to as monocots or grasses. The most important crop species are monocots. All the major cereals—corn, rice, wheat, barley, oats, and rye—plus several other important edible plants including garlic, onions, palm, and banana, are monocots. Weeds on the other hand are mostly dicots. Although both monocots and dicots use auxins as their major hormones, the exogenous application (application from outside of the organism) of auxins or auxin-like chemicals kills dicots but not monocots. (Auxin-like chemicals work on monocot cells but not whole plants.) It is still not completely clear why this is so, but the most likely explanation is that the two plant types differ in their ability either to take up or degrade exogenously applied auxins. Monocots can thus avoid the effects of applied auxins, but dicots cannot. Animals do not use auxins as hormones, and as a result these chemicals are only moderately toxic to animals, and certainly so in the short term.

2,4-D, one of the most widely used herbicides in the world, is produced by numerous chemical companies and is the main ingredient in more than 1,500 herbicide products. With the enormous commercial success of 2,4-D, many chemical companies initiated research programs to find new herbicides. This has not been easy. Most of these second-generation herbicides are toxic to all plants and as a result can be used only to generally control the growth of vegetation, for example to prepare a new site for a construction project or to clear plants from the sides of highways. One of the herbicides of this type is glyphosate (Roundup), discovered by the agrochemical company Monsanto in 1970.

Glyphosate inhibits an enzyme that all plants must use to synthesize the amino acids phenylalanine, tyrosine, and tryptophan. Inhibiting this enzyme kills plants. These three amino acids are essential for animals, meaning that animals cannot make them and instead acquire them through their diet. Consequently, because animals lack the synthetic biochemical pathway that glyphosate targets, they are largely insensitive to this herbicide. (Glyphosate has lower chronic toxicity to humans than 90 percent of all herbicides.)

Since glyphosate kills all plants, it was not useful in crop protection. However, in the late 1980s and early 1990s, new methods were developed to genetically engineer plants. The resulting genetically engineered plants are called genetically modified organisms or GMOs. Monsanto realized that the GMO technology could be used to create crops that are artificially resistant to glyphosate. For such GMO plants, glyphosate would become a great herbicide, killing all the weeds but not the crop.

Commercially, this was a terrific deal for the manufacturer. Not only could Monsanto now make money selling more glyphosate, it could also make money by selling glyphosate-resistant crop seeds. Glyphosate-resistant soybeans were introduced in 1996, and today more than 80 percent of the soybeans grown in the United States are genetically modified to be glyphosate resistant. Many other similarly genetically modified crops were later introduced, allowing these crops to be used with glyphosate herbicide: corn, soybeans, cotton, and canola oil seed. As a result, the use of herbicides has increased manyfold. Today, total worldwide pesticide sales are almost $250 billion, of which about 20 percent is spent on herbicides.

Fungi that cause plant disease

The most important pathogens for animals are bacteria and viruses. Fungal pathogens play a tiny role in animal disease. The opposite is true for plants. Fungal diseases are the most important plant diseases. For most of human history, little or nothing could be done to control fungal plant diseases.

The most horrifying recent example of the effect of fungal disease on crops was the European potato famine of the mid-1840s. Potatoes are not a European plant. When the conquistadores first arrived in modern-day Peru in the mid-1600s, they found that the Indigenous people had domesticated the potato, a plant native to the Andes Mountains. The conquistadores brought potatoes back to Europe sometime before the end of the sixteenth century, and potatoes were quickly adopted as crops in Spain and the British Isles and later in the rest of Europe. Prior to the seventeenth century, there were no potatoes in Europe. The French had no French fries, the British had no mashed potatoes, and Ashkenazi Jews had no potato latkes for their Hannukah celebrations. But once introduced, potatoes turned out to be a high-value crop: easy to grow, resistant to spoilage, and an inexpensive, highly nutritious food. Although native to the Andes Mountains in South America, potatoes thrived in the European climate.

By pure luck, the conquistadores brought potatoes to Europe without their most significant disease, potato blight, a fungus that scientists call *Phytophthora infestans*. This changed in the 1840s, when potato blight was accidentally introduced into Europe via potato shipments from South America. Potato blight devastated crops throughout Europe. In Ireland, which was heavily dependent on the potato as a monoculture crop, there were one million deaths from starvation. An additional one hundred thousand starvation deaths occurred in Belgium, Prussia, and France. More than 1.5 million starving adults and children left Ireland to seek refuge from the famine in America.

Bordeaux mixture, the first fungicide used to treat plants,

was introduced in 1892. It is composed of hydrated lime, copper sulfate, and water. Although hardly antifungal selective (it is extremely toxic to animals as well as fungi), Bordeaux mixture was adopted to control fungal disease in high-value crops like wine grapes. Other various mixtures of toxic copper, sulfur, and phosphorous acid were also developed and used around this time. In the early twentieth century, a chemist named Henry D. Hooker wrote that no entirely satisfactory fungicides were available and described Bordeaux mixture and other lime-sulfur treatments as being "most disagreeable to handle." Better fungicides were desperately needed, but heavy metal fungicides remained the sole treatment option until the 1940s.

Starting in the 1940s, about a dozen new classes of organic chemical fungicides were developed. These chemicals were an enormous improvement over the earlier noxious and toxic metal fungicides, but they still had issues with animal and environmental toxicity. The problem is that the underlying biochemical processes in fungi differ only slightly from the corresponding processes in plants and animals.

As an example, three major classes of these new fungicides, the morpholines, imidazoles, and triazoles, all block the synthesis of sterols, biochemicals that are essential for the life of plants, animals, and fungi. Sterol biochemistry differs slightly in plants, animals, and fungi, and these fungicides target these biochemical differences, producing a degree of antifungal selectivity. Animals make cholesterol, while plants and fungi make sterols that are similar to cholesterol but biochemically distinct from it. Plants make sitosterol and stigmasterol, while fungi synthesize ergosterol. (In humans, cholesterol comes both from endogenous synthesis and from the diet. It is because plants and mushrooms, which are fungi, do not make cholesterol that a vegetarian diet is recommended for patients with hypercholesterolemia. Not consuming cholesterol in the diet lowers serum cholesterol levels.)

Another class, the benzimidazoles, attacks a cellular structure called the microtubule. The microtubules in fungi differ slightly from the microtubules of animals and plants, so these fungicides have some selective antifungal toxicity, but the selectivity is relative and not absolute. As a result, these chemicals have some acute toxic effects on plants and animals and, as time went on, all of these chemicals were found to produce environmental toxicity.

Insects

Insects are the most important and historically the most feared class of crop pests. In the biblical Book of Exodus, God did not inflict the Egyptians with plagues of weeds or plant disease fungi. Instead, two of his ten plagues were insects: lice and locusts. And for most of human history little could be done against insect pests except for hoping and praying.

Dichlorodiphenyltrichloroethane, commonly known as DDT, is an organochloride chemical first synthesized in the late nineteenth century. It was known only as a chemical reagent until 1939, when Swiss chemist Paul Hermann Müller discovered that DDT was a potent insecticide. In 1948 Müller was awarded the Nobel Prize for this work.

Upon learning that Müller won the Nobel Prize for this, many people react by saying his award represented a great injustice and an abject failure of the Nobel Committee decision making process. Why would you give a Nobel Prize to someone who discovered a horribly toxic, environmentally destructive chemical? But this is our twenty-first-century perspective. Müller, although neither a medical scientist nor a physician, received the Nobel Prize in Physiology or Medicine because DDT killed insects that spread malaria and yellow fever. The use of DDT led to the virtual eradication of malaria in some countries where the disease had been endemic since the beginning of time.

Soon DDT also became widely available for use as an agricultural and household pesticide. DDT is a neurotoxin. It opens sodium channels in the insect's nerve cells and produces the uncontrolled firing of nerves, leading to seizures that kill the insect. While DDT also acts on animal nerves, animal nerves are innately less sensitive to DDT than insect nerves, and in addition, animals can rapidly detoxify DDT. A third feature protecting animals is a structure called the blood-brain barrier, which shields most of the nervous system from foreign substances. While animals are relatively resistant to the acute toxicity of DDT and similar insecticides, it would later be discovered that the real problem with DDT is not its acute toxicity to animals, but rather the long-term effects caused by accumulation in the environment.

The popularity and profitability of DDT set off a race by all the major agricultural chemical companies to discover their own DDT-like insecticides. Soon a large number of very similar competing products, called organochloride insecticides, appeared. Some of the most popular insecticides of this type include dieldrin, lindane, mirex, and heptachlor.

Other companies explored unrelated chemicals for insecticidal activity, leading to the discovery of a new insecticide class, the organophosphates. Organophosphates also target the insect's nervous system, doing so by inhibiting a critical nervous system enzyme, acetylcholinesterase. Similar to DDT, these insecticides kill insects by disrupting signaling in the insect nervous system. Commonly used organophosphates include malathion, parathion, and chlorpyrifos. Surprisingly, few people at the time seemed concerned that the organophosphate insecticides were mechanistically similar to the lethal gases used as chemical warfare agents during World War I.

These insecticides became extremely popular with consumers for several reasons. Because they act on the insect nervous system, they quickly paralyze and kill treated insects. Immediately after application you see lots of dead insects. Another popular feature was the fact that these insecticides are long-acting. DDT has a soil half-life of about a month, meaning that after the end of a three-month growing season more than 10 percent of a single application of the chemical is still present, continuing to contribute to insect control.

Among the insecticides, DDT is particularly persistent in the environment. It is poorly water soluble, so it is extremely long-lived in aquatic environments, with a half-life of as long as 150 years. And because of its poor water solubility, DDT becomes extremely stable once ingested by animals, because DDT is stored in fat. In the human body DDT has a half-life of up to ten years.

Especially problematical, DDT shows extreme high accumulation in apex food chain species, especially predatory birds. Each animal in the food chain takes up DDT and stores it in its fat. When that animal is eaten by another animal, its DDT is transferred to the consumer. After multiple cycles, animals at the top of the food chain will build up extraordinarily high levels of DDT and its toxic metabolites. But it unfortunately took years for this to be recognized as a problem.

Invasive Species

In the late 1930s a cargo ship was being unloaded in Mobile, Alabama. In the hold along with the cargo was a stowaway, *Solenopsis invicta*, the South American red fire ant. These fire ants came ashore and thrived in their new home, quickly traveling out from Mobile. Within fifteen years the ants had spread as far north as Tennessee and as far west as Texas. The red fire ant is a terrifying insect pest, a destructive creature unlike any of the insects native to North America.

Red fire ants sting people, producing a painful burning sensation and then leave a smallpox-like pustule at the site of the bite. The ants are voracious eaters and will consume almost anything, including tree bark and other insects including termites, not to mention important cereal crops like wheat and sorghum. They are extremely aggressive. They kill fledgling birds and young sea turtles and have been known to kill baby deer. They construct rigid mounds in farmers' fields that damage harvesting equipment. And if a farmer should be so unlucky as to accidentally disturb a colony, thousands of ants are instantly dispatched to attack the intruder.

In the early 1950s the US Department of Agriculture (USDA) reacted to the fire ant threat exactly as might be expected. It declared all-out chemical warfare against the fire ant.

The USDA started by spraying heptachlor and dieldrin over millions of acres of farmland, killing countless wild birds and large numbers of fish, cows, cats, and dogs. When this strategy turned out to be ineffective in controlling the fire ant, the USDA doubled down and, without much of a scientific basis, started a second chemical warfare campaign, this time using mirex. More than fourteen million acres were sprayed with mirex. In the end, the efforts of the USDA did almost nothing to slow the fire ant invasion, and there is substantial evidence that the USDA's campaign may have actually helped the fire ant to spread by killing off less hardy native ants that were competing with the fire ant for food and habitat.

Academic Scientific Community

One might ask, where were the academic scientists in all this? Why were they not mounting opposition the unproven USDA spraying? Unfortunately, the scientists who were knowledgeable about environmental issues were having their own problems at this time. Modern biologists belong to two distinct schools of thought: the naturalists and the experimentalists. The structure of DNA was published in 1953 by James Watson and Francis Crick. With this powerful new tool, experimental biological scientists quickly achieved the upper hand over their naturalist ecology colleagues.

James Watson became a Harvard professor in 1956. E. O. Wilson, one of the most famous naturalist biologists of our era, was hired by Harvard in the same year. The two scientists quickly became engaged in a tremendous turf battle within the Harvard Biology Department. Similar battles were going on at universities all across the country. The egotistical Watson argued that experimental, molecular science was the only way forward to understand the living world. He maintained that all the naturalists were doing was specimen collecting, the work of hobbyists, not real scientists.

Wilson was a brilliant biologist but ended up having to fight for his academic survival. After a stream of attacks by Watson, Wilson took to calling Watson "the Caligula of biology." Wilson soon received an attractive offer from Stanford University to leave Harvard. To counter Stanford's offer, Harvard granted Wilson early tenure, which had not yet been granted to Watson. Members of the department say that upon hearing he had been bested by Wilson in receiving early tenure, Watson stomped through the halls of the biology building loudly cursing and berating Wilson. Eventually things became so bad between the experimental biologists and the naturalist ecology biologists that Harvard decided that the only way to bring peace was to split Harvard's biology department into two.

I started graduate school in biology at Princeton in 1972. By this time the biological civil war had died down, in no small part because the naturalists were showing that they could also do experiments and not just make observations. One of my Princeton professors was Robert MacArthur, a field biologist who had just completed a ground-breaking study with E. O. Wilson. When the fire ant invaded the United States, there were no scientific principles that could predict how and why the ants would spread. Wilson and MacArthur therefore set about to determine the laws guiding this phenomenon. The result of their collaboration was a 1967 book titled *The Theory of Island Biogeography*.

Wilson and MacArthur started by constructing a mathematical model to describe why large islands harbor more species than smaller ones and why distant islands have fewer species than similar-sized islands situated near to a mainland. This problem had long stymied biologists, including the famous discoverer of evolution, Charles Darwin. MacArthur and Wilson's model proposed that the key was to determine the rate at which new species immigrate to an island versus the rate at which island species die out. This was a simpler problem than explaining the invasion of the fire ant but one whose solution could provide great insights into the US fire ant invasion. Wilson then came up with the experimental plan to test the model.

There are a series of cays, clumps of mangrove trees, just north of Key West in Florida that are so small (about forty feet in diameter) that they have only a tiny number of breeding animals on them, just a few insects, spiders, and an occasional wood louse. Wilson persuaded the National Park Service to let him fumigate six of these cays and then had one of his graduate students monitor what then happened after the cays had been stripped of all animal life. The results showed that their model was correct. The cays closest to the shore were quickly recolonized and species diversity soon rose and then leveled off. Recolonization of the more distant cays took longer, and the eventual number of resident species was lower on the most distant tiny island, exactly as the model predicted. Naturalists were now doing rigorous experimental science just like their molecular experimentalist colleagues. Many scientists cite *Island Biogeography* as being one of the most influential texts on ecology and evolution.

As a Princeton graduate student, I had the choice of specializing in naturalist ecological studies or molecular biology studies. I was tempted to choose the naturalist track. Students on the naturalist ecology track all took a field course in which the faculty and a group of graduate students spent a month in the jungles of South America performing and learning field research. It seemed like it would be more fun to do research in beautiful natural settings rather than in a cold, sterile laboratory.

But one of my classmates took the field course, became overly energetic in swinging his machete, and cut a huge gash in his leg. The faculty and other graduate students had to improvise a litter, and it took two days to carry my bleeding classmate out of the jungle. At that point I decided to study molecular biology and ended up carrying out radiotracer experiments for my doctoral thesis, working with radioactive isotopes of hydrogen, carbon, iodine, and phosphorous. It seemed safer at the time.

The *Island Biogeography* research was completed only five years after the publication of *Silent Spring*. With academic scientists who had the relevant expertise mostly being distracted by their own issues, it fell to Rachel Carson, who had no PhD and was not a practicing scientist, to sound the DDT alarm. In a way, it was extremely fortunate that the person who raised this issue was a writer and not a scientist. Scientific papers can be dry and hard to read. Had the case against pesticides been based solely on scientific publications, it would certainly not have had a broad impact. But Carson, who was an outstanding and inspiring writer, knew how to capture readers' interest. And she also engaged with practicing scientists to make sure they scrutinized her writing for correctness and accuracy.

Carson became focused on the issue of pesticide toxicity in 1958 after receiving a letter from a friend, in which she described the death of birds on her coastal property in Duxbury, Massachusetts, after the aerial spraying of DDT for a mosquito control campaign. This started Carson on a four-year project that culminated with the publication of *Silent Spring* in 1962. She started by collecting examples of environmental damage caused by DDT. She soon connected with a large number of academic scientists whose work had documented an array of physiological and environmental effects that were produced by pesticides. Through her years of work at the United States Fish and Wildlife Service, Carson had developed good connections with many government scientists, who provided her access to their own supportive pesticide studies. Further backing came from medical scientists working at the National Cancer Institute, whose studies showed that many pesticides were powerful carcinogens.

Silent Spring

While some academic scientists at the time were concerned about the possibility that pesticides could be doing significant harm, other scientists considered the dangers to be largely unproven and wouldn't take a position against pesticides until provided with conclusive evidence of causality. The publication of *Silent Spring* ignited a firestorm of controversy.

The chemical industry immediately mounted strong opposition. DuPont, a major manufacturer of DDT and 2,4-D, and Velsicol Chemical Company, the manufacturer of chlordane and heptachlor, were among the first to respond. Velsicol threatened legal action against the publisher, Houghton Mifflin. Brochures and articles were published by chemical companies promoting and defending pesticide use.

Robert White-Stevens, a biochemist at American Cyanamid, wrote, "If man were to follow the teachings of Miss Carson, we would return to the Dark Ages, and the insects and diseases and vermin would once again inherit the earth." Carson was also attacked on the basis of her personal characteristics, scientific credentials, and training in marine biology rather than biochemistry. Former US secretary of agriculture Ezra Taft Benson argued that Carson was "probably a Communist" because she was unmarried despite being physically attractive. Monsanto published a parody of *Silent Spring* titled *The Desolate Year,* which predicted that a reign of famine and disease would soon descend if pesticides were banned.

But the chemical industry's strident opposition served only to raise the public awareness of Carson's ideas and ironically helped to make the case against unfettered use of pesticides. One consequence was the Environmental Protection Agency (EPA), which was created in 1970 by the Nixon administration after almost ten years of campaigning by pro-environmental lobbyists. In the decades following its creation, the EPA has succeeded in implementing numerous policies to significantly improve the environment.

For example, ozone in the stratosphere filters out harmful ultraviolet rays. When released into the atmosphere, chlorofluorocarbon refrigerant chemicals such as Freon break down ozone. By the 1980s, after many years of their use, leaking chlorofluorocarbons had produced a hole in the ozone layer over Antarctica. Congress gave the EPA the power to enforce a ban on chlorofluorocarbons to protect the ozone layer. Together with the efforts of environmental regulators worldwide, this helped the hole to begin to repair itself and, as it later became known that chlorofluorocarbons contribute to global warming, the ban has had a knock-on effect of decreasing the rate of climate change.

By the 1970s America's rivers had been heavily polluted with toxic industrial chemicals and untreated sewage. The 1972 Clean Water Act directed the EPA to clean America's waters and provided billions of dollars in funding. Since then, the waters around major American cities have been dramatically improved. When you fly into Boston's Logan Airport one of the first things you see are the giant white eggs on Deere Island, the site of Boston's sewage treatment plant, which was completed in the year 2000. In the twenty years since its completion, the sewage plant and companion efforts have turned Boston harbor from one of the dirtiest in the world into one of the cleanest.

In 1985 the EPA showed that as many as five thousand people were dying annually from lead-related heart disease. At this time lead was routinely formulated into all sorts of commonly used products, including paint and gasoline. Despite determined objections from industry, the commercial use of lead is now virtually illegal, with concomitant dramatic improvements in public health.

Smog choked many American cities in the 1970s and 1980s. The EPA mandated decreases in the amounts of particulate matter and chemicals in the air, a principal cause of asthma. As a result of the enforcement of EPA regulations, emissions from vehicles and factory and power plant smokestacks declined, even as Americans drove more miles per year and consumed more electricity. This effort resulted in 165,000 fewer annual asthma deaths and a decrease of 1.7 million asthma cases between 1990 and 2010.

For most of the twentieth century, American manufacturing plants legally dumped tons of toxic chemicals into the environment. The 1976 Resource Conservation and the 1976 Recovery and Toxic Substances Control acts gave the EPA the authority to enforce the cleaning of numerous toxic industrial sites, commonly called Superfund sites. A total of 1,344 such sites were identified by the EPA, ranging from Love Canal in New York to the Hanford Department of Energy site in Washington. Of these 1,344 Superfund sites, 413 have now been cleaned up and removed from the list.

Acid rain is caused by emissions of sulfur dioxide and nitrogen oxide into the atmosphere, which react with the water molecules in the air to produce acids. Acid rain damages forests, pollutes fresh water and soil, and kills microbes, insects, and aquatic life forms. It also causes paint to peel, corrodes steel structures such as bridges, and weathers stone buildings and statues. In addition, it has deleterious effects on human health.

In 1990 the US Congress passed amendments to the Clean Air Act designed to control emissions of sulfur dioxide and nitrogen oxides through a cap-and-trade system. The program, administered by the EPA, has successfully reduced sulfur dioxide emissions by 40 percent, and acid rain levels by 65 percent relative to 1976. Enforcement of the 1974 Safe Drinking Water Act by the EPA has significantly improved the quality of American drinking water.

And, most directly influenced by *Silent Spring*, the Environmental Defense Fund (EDF) was founded in 1967, just five years after its publication. The EDF petitioned the government for a ban on DDT and filed lawsuits. In 1971 the US District Court of Appeals ordered the EPA to begin the deregistration procedure for DDT. In the summer of 1972, after seven months of hearings, William Ruckelshaus, the first EPA administrator, announced that the EPA would prohibit DDT use in the United States except for emergency situations.

DDT is particularly toxic to the bald eagle, an American symbol and a species at the top of the food chain. While DDT does not kill adult birds, it interferes with their calcium metabolism. Eggshells are made from calcium carbonate, and DDT causes bald eagles to lay eggs with thin and fragile shells. These eggs break when the eagles incubate them, resulting in the death of the chicks. The banning of DDT was the capstone of decades of legislation to protect the national bird of the United States.

It is estimated that in the early eighteenth century there were 300,000 to 500,000 bald eagles in the United States. By the 1950s there were only 412 nesting pairs in the forty-eight contiguous states. The species was first protected in the US and Canada by the 1918 Migratory Bird Treaty. The Bald and Golden Eagle Protection Act, approved by Congress in 1940, prohibited commercial trapping and killing of the birds. The bald eagle was declared an endangered species in the US in 1967. But the most significant legislative action was the 1972 banning of DDT. The bald eagle population rebounded, reaching 100,000 individuals by the early 1980s. Today there are well over 300,000 bald eagles in the lower forty-eight states.

In fairness, even some strong EPA supporters have found the EPA at times to be overzealous in enforcing their regulations. In the late 1990s, I had an environmental toxicologist colleague who had developed ultrasensitive methods to detect the presence of toxins in environmental samples. One day he was contacted by Environment Canada, the Canadian EPA, which reported that they had found the carcass of a bald eagle but were unable to determine what had killed it. They asked him if he could help them determine whether the eagle had been killed by an environmental toxin.

My colleague said that he would of course be happy to try to help them. A few days later, a freezer chest arrived at his laboratory. As it had gone through US customs, it had a customs tag identifying the contents: bald eagle parts. My colleague immediately started to analyze the sample. Late that day he went home for dinner. When he returned to his lab the next morning to continue his work, FBI agents were waiting to arrest him. They told him that he had violated the Bald Eagle Act, which states that no American citizen, with the exception of certain Native American groups for use in their tribal ceremonies, was allowed to possess any part, including so much as a talon, of a bald eagle.

They asked him what he was doing with the carcass. He replied that he was analyzing it for toxin residues and proudly told the FBI that he had already started to examine the contents of the eagle's stomach and had determined that the eagle had eaten a bird just before it died. Based on his work, he told the FBI that he had identified the species of the bird the eagle had eaten. The FBI now told my colleague that he had in addition violated the Migratory Bird Act by processing bird parts prohibited by the Migratory Bird Act. He was in big trouble.

Fortunately, Environment Canada swiftly came to my colleague's defense. They explained the situation to the FBI and all charges were dropped quickly. My colleague even received an apology from the US government. But for several months thereafter, he entered his lab every morning in a state of high anxiety, not knowing what might await him.

Today we as a society remain far from a consensus on the need for, or benefits from, environmental protection legislation. In December 2016 President Donald Trump nominated Scott Pruitt to become the EPA administrator. Pruitt's nomination was confirmed by the Senate in February 2017. Prior to taking the EPA position, Pruitt had served as the attorney general for the state of Oklahoma, in which role he had sued the EPA at least fourteen times, including efforts to block the anti–climate change Clean Power Plan, to oppose limits on mercury and ozone pollution, to oppose the Cross-State Air Pollution and the Clean Water Rules, and to contest regulations on methane emissions.

As the EPA administrator, Pruitt worked to cut the EPA budget by 24 percent and reduce staffing by 20 percent. He fired scientists from the agency's eighteen-member Board of Scientific Counselors, removed several scientists from EPA advisory panels, and forbade any scientist who received a grant from the EPA from serving on those panels. Less than a year after Pruitt's taking office, seven hundred staff had left EPA, including more than two hundred scientists.

In his first year in office, Pruitt removed, relaxed, or delayed sixty-seven environmental rules. In October 2017 Pruitt issued a notice of proposed rule making to repeal the Clean Power Plan, and in January 2018, he finalized the repeal of the Clean Water Rule. He postponed new effluent guidelines on power plants and regulations of toxins: methylene chloride, N-methylpyrrolidone, and trichloroethylene. Pruitt called for relaxing emission standards for new cars and the repeal of heightened emission standards for "glider trucks." Pruitt refused to ban the organophosphate insecticide chlorpyrifos despite prior findings by EPA scientists that they were unable to find any level of chlorpyrifos exposure that was safe. He also attempted to delay the imposition of an EPA regulation on methane leaks from oil and gas wells despite the well-established fact that methane is eighty times more powerful than CO_2 in trapping heat in Earth's atmosphere.

In 2022, almost fifty years after the publication of *Silent Spring*, in the case of *West Virginia v. Environmental Protection Agency*, the United States Supreme Court decided that the EPA does not have the authority to regulate the carbon emissions of power plants. This ruling reduced the EPA's overall power to regulate carbon dioxide emissions and will almost certainly accelerate the deleterious effects of these emissions on climate change.

Among many ongoing environmental problems, a study published in the fall of 2019 showed that there are now nearly 30 percent fewer birds in North America than there were just fifty years ago. This loss has occurred pretty uniformly across all bird species and types. There is no clear explanation for what caused this decline. And if no one finds out why and figures out how to stop it, our species' decline could potentially be next.

Carson died from breast cancer in 1964 at the age of fifty-six, just two years after the publication of *Silent Spring*. She never knew that significant environmental protections would be put in place as result of its publication. Ironically, following her death, numerous scientific studies have shown that many synthetic substances used worldwide in agriculture, industry, and consumer products can act as endocrine disrupters. These chemicals, now called xenoestrogens, interact with estrogen receptors or their estrogen signaling pathways, or both. Xenoestrogens produce many adverse human health effects and appear to serve an important role in the current increased incidence of breast cancer in the United States and numerous other countries.

Unlike some other scientists in this book, Carson was not just fighting the scientific establishment. In the struggle to have her ideas accepted, she faced challenges on many sides. Her obvious allies in the academic community were unfortunately too busy fighting battles of their own to help her. There was powerful opposition from the chemical industry, and other industrial sectors joined the chemical industry in their opposition. Some political groups also opposed her ideas. And popular opinion has been and continues to be divided regarding her ideas.

9 | Stanley Prusiner
Scientist who discovered prions

Neurological Diseases

Neurological diseases generally appear at random, seemingly by chance, with no clear pattern or apparent direct cause. It is almost never exactly clear why the disease appears in one person and not another. Common examples include amyotrophic lateral sclerosis (also called ALS or Lou Gehrig's disease), Alzheimer's disease, epilepsy, Huntington's disease, multiple sclerosis, and Parkinson's disease. Because of their unpredictable appearance, these diseases are called sporadic.

More rarely, some neurological diseases can instead be familial, caused by the inheritance of a defective gene from one or both parents. About 10 percent of the cases of Alzheimer's disease, ALS, and Parkinson's disease are familial. The singer-songwriter Woody Guthrie died from Huntington's disease, an inherited neurological disease that is exclusively familial.

Historically, the one disease mechanism that was never seen in neurological diseases is an infection. People do not catch neurological diseases from one another. You don't tell your kids to stay away from grandma so they won't catch Alzheimer's. But over the last couple of centuries, medical scientists started to become aware of a group of extremely rare neurological diseases that are caused by an infection. The first to be identified was scrapie, which afflicts sheep.

The eighteenth-century invention of mechanical looms in England greatly increased cloth production and, as a result, the demand for wool. In response, Spanish shepherds started breeding programs to increase the number of animals in their flocks as well as the amount of wool each animal could produce. Soon thereafter a new disease, scrapie, named after the characteristic symptom the disease produced, appeared in some of these inbred animals. Affected sheep compulsively scrape themselves against rocks, trees, or fences. Later in the progression of the disease, more severe neurological symptoms appear, eventually leading to convulsive collapse and death.

It cannot be known with certainty whether scrapie had previously been present in sheep and not noticed or whether the recent breeding program had by bad luck greatly increased the incidence of the disease by including scrapie animals. But veterinarians soon took notice and began to study it.

In 1759 German veterinary scientist Johann George Leopoldt wrote:

Some sheep also suffer from scrapie, which can be identified by the fact that affected animals lie down, bite at their feet and legs, rub their backs against posts, fail to thrive, stop feeding and finally become lame. They drag themselves along, gradually become emaciated and die. Scrapie is incurable. The best solution, therefore, is for a shepherd who notices that one of his animals is suffering from scrapie, to dispose of it quickly and slaughter it away from the manorial lands, for consumption by the servants of the nobleman. A shepherd must isolate such an animal from healthy stock immediately because it is infectious and can cause serious harm to the flock.

Leopoldt's conclusion that scrapie is infectious was not much more than a guess, and a lucky one at that. The first experiments to prove that scrapie was an infectious disease were carried out by French scientists a century and a half later. They attempted to transmit the disease to healthy sheep by exposing them to scrapie animals or by inoculating healthy animals with brain material or blood from affected animals. After several months of observation, no disease transmission was detected.

This experiment failed because scientists at the time were unaware of the extraordinarily long incubation period for scrapie. Scrapie symptoms do not appear until at least eighteen months after infection. When I first learned about this disease, I was taught incorrectly that scrapie was caused by a slow virus, a special type of virus that for unknown reasons takes years to produce any symptoms. But the truth is that, in 1936, more than 150 years after Leopoldt's experiment, two French scientists showed that they were able to infect healthy sheep by inoculating them with brain or spinal cord material from an infected animal and then waiting well over a year for symptoms to appear. This study proved that scrapie was infectious but did not identify what kind of strange pathogen caused scrapie.

In 1986 a similar disease, bovine spongiform encephalopathy, or mad cow disease as it is much more commonly known, was first reported in the United Kingdom. Cattle feed in Europe was customarily being supplemented with meat and bone meal produced from leftovers of the slaughtering process and from the carcasses of sick and injured animals, including sheep. It is believed that meat and bone meal from scrapie-infected sheep caused the disease to jump from sheep to cattle. Subsequently, the disease jumped again, this time to humans. Since 1986 more than two hundred people have died from a neurological disease named variant Creutzfeldt–Jakob disease, the human form of scrapie, believed to have been caused by people eating tainted beef.

Kuru

In the early 1950s a disease similar to scrapie was first reported in humans. This was kuru, a rare, incurable, and invariably fatal neurodegenerative illness. Kuru was discovered by Australian officers patrolling the Eastern New Guinea Highlands of Papua, New Guinea, who found large numbers of sick kuru patients among the Fore people.

The term *kuru* comes from the Fore word *kúru*, meaning "trembling." The most striking symptom of kuru is body tremors. Kuru is also called the "laughing sickness" because patients suffering from the disease often have sudden outbursts of inappropriate laughter. Years of subsequent scientific study have now fully characterized the symptoms, the brain pathology, and disease progression in kuru.

After patients contract the disease, they have no symptoms for about a dozen years. When symptoms finally do appear, they progress rapidly and are unrelenting, leading to death one to two years later. At first patients have an unsteady gait, tremors, and difficulty pronouncing words. Soon they can no longer walk. They show emotional instability and depression. Finally, they are unable to sit, speak, or swallow food. They die from malnutrition or pneumonia or the infection of ulcerated wounds that spontaneously appear over the entire body.

The Australians were alarmed that there seemed to be a rapid rise in the number of cases of kuru, which soon reached epidemic proportions. Most diseases do not appear in epidemics. The number of people who have a stroke or heart disease or arthritis or Alzheimer's disease remains pretty constant year after year and presumably will continue to do so until someone comes up with a great cure or preventive treatment. Only infectious diseases like COVID-19, the plague, cholera, or AIDS appear as epidemics. Therefore, the appearance of the kuru as an epidemic implied that this was an infectious disease. The problem was that everyone at the time knew human neurological diseases were not supposed to be infectious. If kuru was, it would be a very special case, requiring some extremely creative and ingenious thinking to figure out what was going on.

Proving that kuru was an infectious disease and then figuring out what infection caused it was a daunting task. The earliest studies suggested that it had a long incubation period and that patients could have no symptoms for perhaps ten or more years after they contracted the disease. No one was going to remember what they were doing ten years earlier when, unbeknownst to them, they caught kuru. Were they stung by a mosquito? Did someone manifesting kuru cough or sneeze on them? Did they eat some kind of spoiled food that carried the kuru germ? The vital clue came when scientists realized that the doctors treating the kuru patients never got kuru. Doctors are at risk when treating patients in epidemics, so if kuru was transmitted by coughing or sneezing or something in the food or from a mosquito bite or anything like that, some of the doctors would have caught it. Therefore, it had to be caused by something the Fore people were exposed to, but that their doctors, who lived closely among the Fore, were not.

The best supposition was that the infection was caused by something associated with the Fore people's unique and bizarre funeral practices. In Fore culture, corpses were initially buried for a few days and then exhumed once the corpses become infested with maggots. To show respect and as part of mourning, the corpse was then dismembered, cooked, and consumed by family members and served with the maggots as a side dish.

People developed kuru about ten years after a relative had died from the disease. Women and children consumed the brain and viscera of the corpse while men preferred to consume the muscles. Women and children developed kuru much more commonly than the men, suggesting that the infectious agent was mainly present in the brains of the corpses. But what was the disease agent in the brain? To find it, a task once again made challenging by the disease's extremely long incubation period, scientists set out to examine brain material from people who had died from kuru.

It was a virologist named Daniel Carleton Gajdusek who provided proof of a kuru-causing germ in the diseased brains. To do so, Gajdusek transferred brain material from deceased kuru patients into the brains of chimpanzees by drilling holes into chimps' heads and then introducing pureed kuru patient brain matter into the cerebellum. Two years later the chimpanzees started to develop kuru. This was a great finding, the reward for great patience: kuru was spread by a germ. Another reward for Gajdusek was receiving the Nobel Prize in Physiology or Medicine in 1976 for this work.

All germs are tiny, microscopic organisms. The germs that cause infectious diseases are of three types. The largest, which are still far too small to be seen without the aid of a powerful microscope, are the fungi. Fungal diseases are generally minor in humans but can be very bothersome. Examples include athlete's foot, ringworm, dandruff, and Candidal vaginitis.

The next smallest germs in terms of size are the bacteria. Bacteria are just big enough to be seen with a standard light microscope. They produce many serious diseases ranging from strep throat to rheumatic fever, Lyme disease, impetigo, syphilis, gonorrhea, and tuberculosis.

The tiniest germs are the viruses. They are too small to be seen with a conventional light microscope but can be visualized with much more powerful electron microscopes. Viruses produce diseases ranging from influenza to childhood diseases like measles, mumps, and chicken pox to AIDS and COVID-19. Viruses are so tiny and so simple that scientists question whether they are actually alive. Viruses can sit lifeless and inert in a test tube for decades, or even centuries, and then come to life when they encounter a host cell, hijacking it to start producing large quantities of virus.

Gajdusek asked which type of germ caused kuru. Was it a fungus, was it a bacterium, or was it a virus? He concluded it was likely a virus, but a lot of things we know about viruses did not fit the kuru germ, so he called it an "unconventional virus." Normal viruses act fast, but kuru took years to develop. Normal viruses stimulate the immune system, but kuru did not. Also, although viruses are very simple, they must contain a nucleic acid—DNA or RNA. The unconventional kuru virus appeared to have neither DNA nor RNA, but Gajdusek believed it was just a matter of time until the kuru nucleic acid was detected. Probably RNA, he speculated. An incomplete story, but his great scientific detective work richly deserved the Nobel Prize.

Following his discovery, many scientists went in search of the pesky nucleic acid that must be part of the kuru virus but seemed able somehow to conceal itself from the prying eyes of investigating scientists. Despite considerable effort, no one was able to do so. One scientist, however, thought perhaps Gajdusek and the many others couldn't find the nucleic acid because it wasn't actually there. Therefore, there had to be another explanation. That scientist was Stanley B. Prusiner.

Early Life and Medical Training

Stanley Prusiner was born in 1942 in Des Moines, Iowa, and spent most of his childhood in Cincinnati, Ohio. After high school he attended the University of Pennsylvania, where he majored in chemistry. As an undergraduate student Prusiner worked in the laboratory of Sidney Wolfson, a professor in the department of surgery at the University of Pennsylvania Medical School, on a research project studying hypothermia.

In his Nobel Prize autobiography, Prusiner described his positive experiences in Wolfson's laboratory as the reason he decided to remain at Penn after graduation to pursue a medical degree. Soon he was hooked on medical research, and while other medical students were pursuing clinical rotations, Prusiner took on research projects. This was happening during the Vietnam War. In 1968 as Prusiner was completing his medical degree, almost 300,000 young American men were drafted to fight in Vietnam, and the draft affected decisions made by every student in that era. For medical students there was the Berry Plan, a government program that allowed physicians in training to defer obligatory military service until they had completed medical school and medical residency. Berry Plan participants would then be required to enter the armed forces as physicians following their training.

Of those in the Berry Plan program, an elite group of about a thousand talented young doctors were given the option of joining the National Institutes of Health (NIH) to fulfill their military obligations by being trained as clinical investigators rather than solely as clinical practitioners. Participants in the NIH program humorously called themselves "Yellow Berets." Prusiner was one of these. After a one-year internship at the University of California, San Francisco (UCSF), Prusiner started in the NIH program working in the laboratory of Earl Stadtman. There he mastered a broad array of new skills: developing assays, purifying macromolecules, documenting discoveries using varied approaches, and learning how to write clear and convincing scientific manuscripts.

A common next step in Prusiner's training would have been a postdoctoral fellowship, but instead he decided to take a residency in neurology at UCSF. It was a fateful choice. Two months after starting his residency, Prusiner admitted a female patient who was exhibiting progressive loss of memory and had difficulty performing some routine tasks. Prusiner wrote in his Nobel Prize autobiography:

> I was surprised to learn that she was dying of a "slow virus" infection called Creutzfeldt-Jakob disease (CJD) which evoked no response from the body's defenses. Next, I learned that scientists were unsure if a virus was really the cause of CJD since the causative infectious agent had some unusual properties. The amazing properties of the presumed causative "slow virus" captivated my imagination, and I began to think that defining the molecular structure of this elusive agent might be a wonderful research project. The more that I read

about CJD and the seemingly related diseases—kuru of the Fore people of New Guinea and scrapie of sheep—the more captivated I became.

Understanding what caused this obscure disease became Prusiner's life work.

Prion Hypothesis

After completing his neurology residency, Prusiner was offered a tenure-track assistant professorship at UCSF. Research studies on the diseases that Prusiner was later to name *prion diseases* were conducted using disease models in mice and hamsters. Because disease symptoms develop so slowly, each experiment takes most of a year to complete. This would be a terrible project for a young faculty member, who would need to publish a large number of high-quality research papers in order to be granted tenure. In order to produce enough papers to receive tenure, many professors at the time worked with cheap, rapidly growing experimental systems such as the bacterium *E. coli*, which divides once every twenty minutes, or yeast, which divides once every one hundred minutes. In addition to the long-term experimental protocols, scrapie research required thousands of expensive hamsters and mice. Obtaining funding for such an enormous endeavor was a challenge in its own right.

Prusiner hedged his bets and started off writing grant proposals on glutamate metabolism, a system he had previously worked on during his early scientific training. In parallel, he pursued the difficult task of obtaining grant funding to work on scrapie. After initial rejection following study section review, a process in which a panel of ten or so expert scientists makes recommendations to the NIH regarding which grant applications to fund, Prusiner set up a collaboration with William Hadlow and Carl Eklund, established scrapie researchers who lent credibility to his applications, and was finally able to obtain the funding for his scrapie research.

Prusiner's initial working hypothesis was that the scrapie agent would turn out to be a small virus. But in experiment after experiment, his preparations of the suspected virus contained protein but no nucleic acid. After Prusiner published a succession of experiments failing to show that the scrapie pathogen contained nucleic acid, the Howard Hughes Medical Institute, which had provided financial support for his research, informed him that they wouldn't renew their support. Shortly thereafter, UCSF told Prusiner that he wouldn't be granted tenure (although this decision was later reversed).

Science does not progress through negative data. Just because you cannot find something does not prove that it's not there. It was going to take an enormous amount of data, outstanding experimental design, and a compelling hypothesis to convince the scientific community that there could be an infectious agent that had no nucleic acid. As recently as the last decade or so of the twentieth century, this finding would be considered by most scientists to be total heresy. Everyone knew that living things needed nucleic acid to replicate and grow. Without nucleic acid, there could be no replication, no multiplication, no life. An organism without nucleic acid was a dead end.

But the more work he did, the more confident Prusiner became that the existence of a nucleic acid–lacking pathogen was the only coherent explanation for diseases like scrapie. In 1982 Prusiner published a major paper outlining the evidence that the agent that caused scrapie and related diseases contained no nucleic acid. In this paper he also proposed hypotheses for how such a pathogen could function. Prusiner coined a new term for his new type of pathogen, *prion*, a portmanteau of the words protein and infection meaning a "proteinaceous infectious particle."

The publication of this paper, "Novel Proteinaceous Infectious Particles Cause Scrapie," ignited an explosion of condemnation of and contempt for Prusiner's thinking. Some investigators were merely skeptical while others were angry. Although journalists generally didn't understand the scientific arguments involved, they found the dispute itself to be newsworthy. Lacking comprehension of the underlying scientific issues, they instead focused on the quarrel among scientists, providing Prusiner's critics with a platform in the popular press where they could launch nasty personal attacks against his ideas.

I remember going to talks by Stanley Prusiner at scientific meetings. Scientists in the audience would voice big problems with Prusiner's hypothesis. They'd look at his data and come up with alternate explanations that required a nucleic acid to be there—the nucleic acid just hadn't yet had been found, they argued. Prusiner never got angry at his critics. He never engaged in heated arguments. He just listened politely to what the other scientists said.

The next year, at the next conference, Prusiner presented new data that addressed the dissenting hypotheses he had heard the year before, data that dismissed his critics' ideas. And at this following meeting there was a new but smaller set of alternate explanations proposed by scientists in the audience. Again, Prusiner never got angry. He never engaged in heated arguments and just listened politely to what the other scientists said.

Year after year, after each conference he returned the following year with new data that eliminated the arguments he had heard the prior year. Eventually, there were no scientists standing up and proposing counterhypotheses. Prusiner had, step by step, eliminated any explanation except for his own heretical idea that the pathogens for diseases like kuru were proteins and that no nucleic acid was involved.

Prusiner's hypothesis provided a highly novel process by which something could live and propagate without nucleic acid. In his model, the scrapie pathogen is a single protein, called PrP^{Sc}. PrP^{Sc} is a misfolded version of an ordinary mammalian protein, PrP^{C}. Very strangely, and despite extensive research, the normal function of PrP^{C} in healthy animals and people remains unknown to this day. Fortunately, not knowing the function of PrP^{C} is not problematical because the disease produced by PrP^{Sc} is due to the gain of two new functions by the misfolded PrP^{Sc} protein. One of these functions is toxicity. PrP^{Sc} becomes toxic to neurons in the central nervous system, producing neural abnormalities and neuron loss, thus accounting for the disease symptoms of scrapie. The second new function is templating. When PrP^{Sc} encounters a molecule of normal PrP^{C} it acts like a mold to convert the normal PrP^{C} into pathogenic PrP^{Sc}. Thus, PrP^{Sc} replicates itself without nucleic acid. With time, after many such encounters, the level of PrP^{Sc} grows and grows until disease symptoms are produced.

Showing this was not easy. One common strategy used by biological scientists to tackle difficult scientific questions is to find a simple system through which they can study their problem. This strategy can eliminate a whole host of potentially extraneous and misdirecting issues and observations from their research, issues that can lead the researcher astray and confound data interpretation. The simple system focuses thinking on a limited data set and thus a small number of possible valid hypotheses. In addition, clean and precise technical manipulations have been developed in such simple systems, allowing precise experimental designs that would be impossible using a complex system. Prusiner and other researchers took this approach to study prions.

The yeast *Saccharomyces cerevisiae* is a single-celled microorganism that is used for the leavening of bread and the fermentation of grains and fruits to produce alcohol. That is basically all it does other than divide and grow. In the mid-1960s reports appeared of yeast strains that were infected with viruses, viruses that later turned out to be prions. One of these infections was called [URE3]. In yeast, the Ure2 protein blocks the uptake of poor nitrogen sources when yeast is grown on a good nitrogen source. Normally, yeast grows efficiently, always utilizing the best nutrient source. But in [URE3] strains, this efficient growth does not occur, and nutrients that are poor sources of nitrogen are utilized instead.

When researchers first encountered [URE3] yeast strains, their first thought was that the [URE3] metabolic effect was caused by a new mutation. But they soon determined that [URE3] was not transmitted like a mutation. It had to be something else. The critical experiment to understand [URE3] employed a yeast technique called cytoduction.

Higher cells, including yeast, have two compartments. One of these is the nucleus, which contains the DNA incorporated into chromosomes. The other compartment is the cytoplasm, which contains pretty much everything else. In the cytoduction experiment, a test yeast strain was crossed with a [URE3] strain. The [URE3] strain can donate either its nucleus or its cytoplasm to the test strain, but not both, and what happened was that [URE3] came into the test strain via the cytoplasm but never the nucleus. Had [URE3] been produced by a mutation, the mutation would be in the DNA, which is in the nucleus. But [URE3] was not transmitted by the nucleus, so therefore [URE3] was not caused by a mutation.

Something in the cytoplasm was producing [URE3]. During my training I was taught that [URE3] was most likely an RNA virus present in the cytoplasm. But what if instead [URE3] was a prion? What was needed was an experiment that could distinguish between a virus and a prion. One difference between the two infectious agents is that a virus needs only a cellular environment to replicate while a prion needs both a cellular environment and a protein that the prion can change into a toxic form via its templating activity. A second cytoduction experiment was therefore designed to test this. Precise alterations in yeast DNA can be easily and conveniently constructed. In this second cytoduction experiment, the gene that produces the normal Ure2 protein was deleted from the test recipient strain. Nothing else was changed.

In this variant experiment, the test recipient yeast strain was unable to make the normal Ure2 protein. Shortly after cytoduction, [URE3] was lost from the cytoplasm. This occurred because in the absence of the normal Ure2 protein, the misfolded [URE3] protein donated in the cytoduction experiment could not replicate itself, and with time it aged, was metabolized, and was lost. A source of normal Ure2 protein to template was essential for making more [URE3] prion. Thus [URE3] is a prion, not a virus.

No other model could explain all the data from this and a huge number of other experiments. Prusiner was correct. In 1997 Prusiner was awarded the Nobel Prize in Physiology or Medicine for his work. He was fifty-five years old and had worked on prions for his entire professional life. As wild and improbable an idea as it was, the notion that something could replicate without nucleic acid turned out to be true. Prusiner's opponents finally gave up and accepted that he was correct.

Prusiner's journey from presenting a supposedly absurd hypothesis to its scientific recognition took fifteen years, an amazingly rapid acceptance for an idea that defied the fundamental conventions of his field. But his journey was by no means easy. Prusiner had to withstand the extreme criticism of his peers for many years. His success was likely due to his methodical strategy, fighting disdain and disparagement with exceptional experimental design and formidable data, all the while maintaining a high level of professionalism in his interactions with colleagues.

10 | Amedeo Avogadro
Determined how many molecules are present in defined samples of matter

Atomic Theory

Humans have long wondered what matter is composed of. The idea that matter might be made up of tiny, invisible, discrete particles was pondered in antiquity. The Greek word *atomos*, used to denote the basic unit of matter and meaning "uncuttable," was first proposed by the philosophers Leucippus and Democritus in the fourth century BC to describe the basic building blocks of our physical world. Later Greek and Roman philosophers expanded this idea.

These ancient atomic theorists proposed that nature consists of two fundamental principles: atom and void. They argued that all the diverse substances in the world are the result of various arrangements of atoms in different shapes, groupings, and positions—which is not far from being correct. But this was really just speculation, an interesting idea unsupported by evidence, made by some very smart people, but people who were sometimes correct and also sometimes terribly wrong.

The concept that matter was composed of tiny particles was forgotten for many centuries and then resurfaced in medieval Europe, when scholars discovered it in ancient Greek texts. Christian theologians opposed the idea, because it implied that matter was the result of chance associations of invisible particles, which seemed to repudiate the Christian belief that the universe had been created by God, who imposed his own special order upon it. In the seventeenth century, prominent scientists Robert Boyle and Isaac Newton supported the theory but with little or no corroborating scientific evidence.

Later, near the end of the eighteenth century, this idea came to be used to explain a series of important experimental observations made by the English scientist John Dalton. As was described earlier in the chapter on Max Planck, Dalton studied chemical oxides, compounds made from oxygen and three other elements: tin, iron, and nitrogen. He found that there were multiple ways in which each of the three elements could combine with oxygen: tin and iron each combined with oxygen in two ways and nitrogen in three ways. But for a given fixed amount of tin, iron, or nitrogen, in every case oxygen combined with the element in whole number ratios.

For tin, one of the oxides (SnO) is composed of 100 parts tin and 13.5 parts oxygen while the other (SnO_2) is composed of 100 parts tin and 27 parts oxygen: an oxygen ratio of 1:2. For iron, one of the oxides (FeO) is composed of 100 parts iron and 28 parts oxygen while the other (Fe_2O_3, also called rust) is composed of 100 parts iron and 42 parts oxygen: an oxygen ratio of 2:3. And for nitrogen, one of the oxides (N_2O) is composed of 140 parts nitrogen and 80 parts oxygen, another (NO) is composed of 140 parts nitrogen and 160 parts oxygen, and the third (NO_2) is composed of 140 parts nitrogen and 320 parts oxygen: oxygen ratios of 1:2:4.

Dalton wrote that these whole number ratios were best explained if discrete, individual particles of these elements were reacting with discrete individual particles of oxygen. Borrowing from the Greek philosophers, Dalton called these particles "atoms," using the term to refer to the basic particles comprising *any* chemical substance.

Today we use the term *atom* to refer only to the particles that make up the chemicals that we now call elements. Elements are fundamental types of matter and cannot be further separated into component chemical parts (although they can be separated into subatomic particles like protons, neutrons, and electrons). Today, chemicals that are made up of combinations of elements and can thus be further reduced to component element chemical parts are known as "molecules" or "compounds." But in the early nineteenth century, the terms "molecule" and "atom" were used interchangeably, and this lack of distinction produced serious problems for the formulation of atomic theory.

William of Ockham, a fourteenth-century English Franciscan friar and philosopher, proposed an idea that has become known as Occam's razor. Occam's razor states, "All things being equal, the simplest solution is commonly the best one." Occam's razor is commonly correct, but not always. Dalton proposed that atoms existed as single particles in all chemical reactions, although he never imagined that following a chemical reaction, the atoms involved in the reaction would somehow be physically linked. He argued for the absence of linkage as the simplest explanation.

In 1809 French chemist Joseph Gay-Lussac reported that when two liters of hydrogen gas react with one liter of oxygen gas, they form two liters of gaseous water (water vapor): a two-to-one ratio of hydrogen to oxygen. Today we know water contains two hydrogen atoms for every oxygen atom and as a result we write the formula for water as H_2O. But incorrectly thinking that the simplest solution was likely to be the correct one, in this case

Dalton postulated that water was made up of one hydrogen atom and one oxygen atom. The simplest chemical formed from any two elements would be one atom of each. He therefore wrote the formula for water as HO. To reconcile this idea with Gay-Lussac's observations, Dalton said that there were twice as many atoms in a liter of oxygen than in a liter of hydrogen. Therefore, you needed to use half the size volume of oxygen to produce a one-to-one ratio of oxygen to hydrogen atoms.

Amedeo Avogadro

Amedeo Avogadro was born into a noble family in Turin in 1776. His father, Filippo, was a magistrate and senator who carried the title of count. His mother, Anna Vercellone of Biella, was a noble-woman. Avogadro completed studies of ecclesiastical law at the University of Turin in 1796. Studying to become a church lawyer and theologian would not appear to be an auspicious start for a career in science, but several years after finishing his ecclesiastical law studies Avogadro switched to physics and mathematics (physics at the time was called positive philosophy).

In 1811, Avogadro published a brilliant and astonishingly innovative article in the *Journal de Physique, de Chimie et d'Histoire Naturelle* (*Journal of Physics, Chemistry, and Natural History*). The paper was titled "Essai d'une manière de determiner les masses relatives des molecules élémentaires des corps, et les proportions selon lesquelles elles entrent dans ces combinaisons" (Essay on a Manner of Determining the Relative Masses of the Elementary Molecules of Bodies and the Proportions by Which They Enter These Combinations).

Working to reconcile all experimental observations up to that point, Avogadro proposed that equal volumes of any two gases, at equal temperatures and pressures, contained equal numbers of particles. That is, that while different gases have different masses, the mass of a gas particle does not affect the volume that it occupies. What then does determine the volume of a gas? According to Avogadro it appeared to be three things. One was the pressure; increasing pressure would force a gas into a smaller volume. A second was temperature; heating causes a gas to expand and occupy a larger volume. The third was the number of particles present; more particles would produce a greater volume.

Thus Avogadro reasoned that if pressure was the same and the temperature was the same, the number of particles also had to be the same. This idea had profound implications. Gay-Lussac's experiment showed that two liters of hydrogen plus one liter of oxygen produce two liters of water vapor instead of three liters of water vapor. Why was one liter of gas lost? It was a puzzle for which neither Dalton nor Gay-Lussac was able to come up with a good explanation. They understood that it was unreasonable to expect water particles to stay closer to one another than oxygen or hydrogen particles, but there seemed to be no other way to explain the observed result.

Avogadro's theory said that the resulting number of water particles was the same as the number of reacting hydrogen particles. The addition of oxygen to the hydrogen did not change the number of particles. He argued that the oxygen was linking to the hydrogen and, as a result, did not increase the number of particles. The hydrogen and oxygen linked together to form a new molecule composed of two different atoms. Avogadro was the first scientist to conceive of a molecule.

But there was still a problem. It seemed that a liter of oxygen contained only half of the number of particles needed to combine with the particles in two liters of hydrogen. Avogadro posited, what if oxygen exists as a molecule itself, composed of two atoms of oxygen linked together? The two atoms in each oxygen molecule could then split, thus providing two particles, enough to combine with all the particles of hydrogen. Today we write the formula for the oxygen molecule, the way oxygen exists in nature, as O_2. The separated oxygen atoms then each combine with a hydrogen particle which, to keep the math straight, themselves exist as a molecule composed of two atoms of hydrogen, H_2 in contemporary nomenclature. Water would therefore be what we now know it to be: H_2O. Avogadro was also the first scientist to realize that elements could exist as molecules rather than as individual atoms.

Avogadro's 1811 publication provided a logical explanation of how elements were able to react and combine with one another and, as he promised, did reconcile all experimental observations up to that point. One would have expected that his ideas would be quickly and eagerly embraced by the scientific community. Instead, Avogadro's findings were largely ignored for the next forty years. Historians have proposed a number of explanations for this rejection.

One school of thought proposes that the lack of acceptance was due to Avogadro's relationship with the scientific establishment of his time. Avogadro did not have broad scientific training, having instead primarily studied ecclesiastical law. As a result, he did not have the support of mentors within the scientific community. There was also the problem of his geographical isolation relative to the scientific community in Europe. Avogadro was Italian, and at the time everyone knew that good science did not come from contemporary Italy. Of course, by way of counterargument, these very factors may have helped Avogadro to develop as an independent thinker and come up with an insight that had eluded the famous establishment chemists of his era.

Second, Avogadro did not work in one primary area of investigation. Instead he studied many unrelated subjects, including the specific heat of gases, affinity of substances for oxygen, the electric force of steam, expansion of mercury when heated, refractory power, and electropositivity. Consequently, in his lifetime Avogadro never became known as an expert in the one subspecialty for which he is now famous, atomic theory. Not being an expert, he was discounted by other scientists as an "amateur." On the other hand, it's known that many other early nineteenth-century scientists worked in a variety of scientific areas during their lifetimes, so arguably Avogadro was not unusual in this regard and this couldn't be the entire explanation.

Third, Avogadro's work was based on theoretical rather than an empirical approaches during an era when experimental evidence was coming to be considered to be of the highest importance. Yet it is also arguable that had Avogadro conducted extensive experiments, the poor scientific instrumentation and flawed general understanding of chemistry of his era might have prevented him from making such critical, far-reaching generalizations as he did.

Last, a fourth school of thought suggests that Avogadro's involvement in politics may have hindered recognition of work. Avogadro was born in the Italian city of Turin and for his entire life lived and worked there. When he was a young man, Turin was part of the kingdom of Piedmont-Sardinia, which was conquered by Napoleon in 1796. The royal family of Piedmont-Sardinia withdrew to Sardinia, which remained free from French control. In 1815, after Napoleon's defeat, the Congress of Vienna reestablished the kingdom of Piedmont-Sardinia, and it came to be ruled by King Victor Emmanuel. Victor Emmanuel abolished all the freedoms granted by Napoleon and instituted a fiercely oppressive rule, reverting to the repressive pre-Napoleonic legal system. He instituted punishing customs barriers to fund the royal treasury, refused to grant a liberal constitution, entrusted education to the Church rather than to more liberal and progressive elements of society, and reinstated laws that discriminated against Jews and Waldensians, an ascetic Protestant sect in the largely Catholic country.

In 1821 Avogadro became active in a revolutionary movement against Victor Emmanuel. Two years later, supporters of the king at the University of Turin forced him out of his professorship, officially stating that they were "very glad to allow this interesting scientist to take a rest from heavy teaching duties in order to be able to give better attention to his researches." It was ten years before Avogadro regained his chair at the university, so, for a critical part of his career, he was without a formal university appointment.

Avogadro died in 1856 at the age of seventy-nine. Two years later, in 1858, an Italian chemist named Stanislao Cannizzaro published a paper titled "Sunto di un corso di Filosofia chimica" (Sketch of a course of chemical philosophy). In this paper Cannizzaro promoted Avogadro's work, arguing for the importance of his insight in distinguishing between the molecular weight of elements and their molecular weights as they appear in nature.

In 1860 the first ever international chemistry meeting was held in Karlsruhe, Germany, as a forum to discuss, and hopefully resolve, a variety of fundamental disagreements that were plaguing chemistry at the time. The Karlsruhe Conference was attended by the most highly regarded chemists of the nineteenth century including August Kekulé, today regarded as the father of modern organic chemistry.

One of the most serious disagreements regarded the correct atomic weights and formulae for common chemicals. As an example, chemists at the time had proposed no fewer than nineteen different formulae for acetic acid, the common organic acid found in vinegar—none of them the correct formula that we use today. Further scientific progress could not be made without the correct understanding of chemical formulae, something that is so fundamental to chemistry.

On the last day of the meeting, reprints of Stanislao Cannizzaro's 1858 paper were distributed to the attendees. Although the meeting ended without an immediate consensus, the Cannizzaro paper made a huge impact on the proceedings. German chemist Lothar Meyer later wrote that upon reading Cannizzaro's reprint, "the scales seemed to fall from my eyes."

With time a consensus was eventually reached through the new understanding of Avogadro's theory. Following the Karlsruhe meeting, the correct atomic weight values of 1 for hydrogen, 12 for carbon, and 16 for oxygen as well as atomic weights for other common elements were finally adopted. It was accepted that Avogadro was correct in describing elements, like hydrogen, nitrogen, and oxygen, as existing as diatomic molecules and not individual atoms.

Avogadro's Number

Avogadro said that if you analyzed any measured sample of matter you would find that it was composed of a precise number of molecules. But how many molecules would that be? One of the most important consequences of Avogadro's work was the development of the concept of the mole (from the German word *Molekül*, meaning "molecule"). The mole is the quantity of a chemical expressed as the number of atoms or molecules of that chemical that are present in a defined mass of chemical matter. The number of particles in a mole is now called Avogadro's number and is defined as 6.022×10^{23}.

But Avogadro did not determine Avogadro's number. He only provided the theory that said such a number should exist. Avogadro's number was determined by scientists working during the half century following his death. Interestingly, this work was performed during a period when those who accepted Avogadro's theory belonged to two opposing camps.

The first group said that Avogadro's theory was a superlative model that explained all data collected up to that point and made excellent testable predictions. However, they also said that Avogadro's particles were a mere mental construct. His particles provided an excellent mental framework through which to think about and solve chemical problems, but they did not exist in reality. The other camp said that the particles actually existed.

The first attempt to derive Avogadro's number was made in 1865 by the Austrian chemist Josef Loschmidt and was based upon his application of the kinetic theory of gases. The kinetic theory of gases states that gases consist of very small particles, negligible in size relative to the volume of the gas. The number of particles is so large that their behavior can be treated statistically. The moving particles collide among themselves and with the walls of the container, but otherwise do not interact and exert no force on one another. The number Loschmidt came up with, called Loschmidt's constant, is 2.69×10^{19} particles per cubic centimeter at standard temperature and pressure (a temperature of

273° Kelvin, or 0° Celsius, and a pressure of 100 kPa). (The metric unit for pressure is the pascal [Pa]. A kilopascal [kPa] is equal to 1,000 pascals—the pressure exerted by a 10 kg mass resting on a 1 cm² area.) Loschmidt's constant is mathematically related to Avogadro's number and is an expression of the exact same chemical concept. Once you correct for volume versus mole, the value determined by Loschmidt is off by only a factor of two from the currently calculated number, but was pretty good for its time.

Fifty years later, a much better determination of Avogadro's number was made based upon a physical phenomenon called Brownian motion. Brownian motion is named after the Scottish botanist Robert Brown. In 1827 Brown reported that while examining grains of pollen suspended in water under a microscope he saw minute particles (now known to be starch organelles) ejected from the pollen grains. Remarkably, the particles moved in a continuous jittery motion. He then repeated the observation with minute particles of inorganic matter, showing that the motion was not life-related. While Brown's observation was clear-cut and simple to reproduce (I did so myself at one point during my early university studies), Brown was not able to come up with an explanation for what produced the random motion of the tiny particles.

Neither was anyone else for many decades. In 1905, no less a scientist than Albert Einstein published an explanation of Brownian motion. Einstein said that the starch organelles were being battered by random collisions with individual water molecules, bouncing them around. Einstein also showed mathematically how Avogadro's number could be derived based on Brownian motion together with a series of physical properties:

$$N_A = (1/[x^2])(RT/3phr)t$$

where N_A is the Avogadro's number, x^2 is the mean square displacement of the particle, R is the gas constant, T is the absolute temperature, h is the viscosity, r is the particle radius, and t is the time interval between measurements of the particle position.

In 1909 the French physicist Jean Baptiste Perrin made physical measurements to determine Avogadro's number based on Einstein's equation. The most difficult physical properties to determine were x^2, the mean square displacement of the particle, and t, the time interval between measurements of the particle position. Perrin used a device called a camera lucida, a state-of-the-art apparatus in 1909, and a stopwatch to determine x^2 and t. He then calculated Avogadro's number, coming up with a much more precise value than Loschmidt. Perrin was awarded the Nobel Prize in Physics in 1926 for providing convincing evidence that atoms and molecules actually exist. But the pursuit of Avogadro's number was not over.

The most effective strategy for the precise determination of Avogadro's number is striking in its cleverness and simplicity. In 1909 the American physicist Robert Millikan determined the charge on an electron, the currently accepted value for which is $1.60217653 \times 10^{-19}$ coulombs per electron. The charge of a mole of electrons, called the Faraday, had been known for some time. The current accepted value of the Faraday is 96,485 coulombs per mole of electrons. Based on these quantities, the strategy to determine Avogadro's number is simple. One takes the Faraday, divides it by the charge of a single electron and the answer is the number of electrons in a mole, Avogadro's number, 6.022×10^{23} particles per mole. This is the value presented in all contemporary chemistry textbooks.

One might ask how such a basic, fundamental, and important

scientific quantity as the number of particles in a mole came to be assigned such an odd number? This occurred because Avogadro's number was determined many years after other important measuring standards had been established and agreed upon. The simplest element is hydrogen. Its molecular weight was therefore set as 1. More complex elements had their atomic weights set relative to hydrogen, 12 for carbon, 16 for oxygen, etc. The quantity to make up a mole was then set at 1 gram for hydrogen, 12 grams for carbon, and 16 grams for oxygen, but what is a gram and where did it come from?

The metric system was developed at the end of the eighteenth century as a byproduct of the French Revolution. In 1790 a French government bureaucrat named Talleyrand proposed that the new French government should develop a new, rational system of measurement based on natural units. Prior to that time a variety of competing arbitrary and imprecise measurement systems had been put into use in various European countries.

The English mile was based on the Roman mile, which was defined as 1,000 paces, measured by every other step. It was also defined as 5,000 feet, a foot being the length of a person's foot, commonly the king's. During the tenth century King Edgar of England defined the standard for length as the distance from the king's nose to the tip of his fingers: the yard, equal to three feet. One hundred years later this standard of measure was modified by King Henry I, who substituted the length of his own arm.

The French metric measurement system was designed to be based on universally accessible natural standards. The new unit of length was the meter, which was set to be equal to one ten-millionth of the length of the distance on the Earth's surface between the equator and the North Pole. Virtually all other commonly used measures were then derived from the meter. But as universally accessible as this is, the metric system is arbitrary in terms of the basic, fundamental composition of the universe. For example, the circumference of Mars is about half that of the Earth. If humanity had evolved on Mars instead of the Earth, the meter would be about half the size we use.

The metric unit of weight is the gram, originally defined as the weight of a volume of water in a cube of one centimeter on a side, a centimeter being one hundredth of a meter. So, a mole of hydrogen weighs one gram, the weight of a small cube of water. But this has nothing to do with the basic deep structure of matter. Despite that, by the mid-nineteenth century chemists were stuck with the gram as defining a mole of hydrogen atoms. As a result, Avogadro's number is defined as an odd number, 6.022×10^{23} particles per mole.

Amedeo Avogadro died in 1856. He wrote his critical scientific paper in 1811, more than four decades prior to his death. No one in the scientific establishment at the time supported his thinking and conclusions. The path from obscurity to recognition began in 1858, almost fifty years after the paper was written and two years after Avogadro's death, when Cannizzaro published his paper explaining and promoting Avogadro's theory. It wasn't until after Cannizzaro's paper was presented at the Karlsruhe Congress two years later that Avogadro's work finally began to achieve scientific recognition. The scientific constant that bears his name, Avogadro's number, was first derived in 1865, almost ten years after his death and more than fifty years after his ideas were first published. The current, correct value for Avogadro's number was determined almost one hundred years after the publication of Avogadro's hypothesis and fifty years after his death. Jean Perrin, one of the scientists who derived Avogadro's number, received the Nobel Prize for showing that atoms and molecules actually exist. Today Avogadro's work is considered to be one of the fundamental pillars of atomic theory. But Avogadro died thinking that he had made no scientific contribution of any special significance.

In a sense, Avogadro is chemistry's version of biology's Gregor Mendel. As with Avogadro, Mendel's work made no impact when it was published in 1865, although in Mendel's case this was due to a mixture of disbelief and the obscurity of his paper. Mendel's work became recognized twenty-five years after its publication, and Avogadro's forty-five years after publication. Both Avogadro's and Mendel's research was recognized through the efforts of other scientists: Cannizzaro for Avogadro and de Vries for Mendel. And for both men, recognition occurred after their deaths, sixteen years after death for Mendel and four years for Avogadro.

11 David Cushman and Miguel Ondetti
Discovered a new standard treatment
for high blood pressure

Hypertension (High Blood Pressure)

Blood pressure is the force produced by the action of the heart as it pumps blood through the circulatory system, causing the pumped blood to push against the walls of the blood vessels. There are two blood pressure numbers. The higher number, systolic blood pressure, is the pressure that is produced during the maximum contraction of the heart muscle. The lower number, diastolic blood pressure, is the pressure produced when the heart is completely relaxed and the chambers of the heart fill with blood.

Today medical textbooks say that normal resting blood pressure is 120 (systolic) over 80 (diastolic). These numbers are the height, in millimeters, that a column of mercury will be pushed up by the pressure of the blood. Although we now call this is normal blood pressure, a medical consensus on what actually constitutes normal blood pressure is surprisingly new.

When I was a student in the 1970s, many textbooks still called high blood pressure that was not the direct consequence of some disease essential hypertension. Today we call it idiopathic hypertension. This type of hypertension was called essential because medical scientists believed, incorrectly, that it was necessary for the health of the patient. Physicians at the time thought some people had high blood pressure because higher blood pressure was needed to adequately perfuse their organs. This higher pressure was thus considered a normal physiological response of those individuals to make sure there was an adequate blood supply to all parts of the body.

In 1931, John Hay, Professor of Medicine at Liverpool University, wrote that "there is some truth in the saying that the greatest danger to a man with a high blood pressure lies in its discovery, because then some fool is certain to try and reduce it." As recently as the 1960s, some physicians were reluctant to lower elevated blood pressure in such patients for fear of harming them.

President Franklin Delano Roosevelt suffered from high blood pressure for most of his life, a condition that worsened steadily over the course of his presidency. The president of the United States of course receives some of the best medical care available from the American medical establishment, state-of-the-art medical care for any medical condition from which they might suffer.

Roosevelt released his medical records during his first election campaign in 1931. These records cited his blood pressure readings as being 140 over 100. After his election Roosevelt selected Admiral Ross McIntire, an eye, ear, and nose specialist, to be his personal physician. In 1937, at the beginning of Roosevelt's second term, the president's blood pressure was 169 over 98. By February 1941 his blood pressure had increased to 188 over 105, although at this point the president had not as yet started to have any cardiac problems. High blood pressure commonly takes many years to produce cardiovascular damage. By January 1944 Roosevelt started to complain about evening headaches. His daughter Anna became concerned and asked the president's physician whether he had checked the president's blood pressure.

McIntire replied that he did not think checking the president's blood pressure was necessary.

During the Yalta Conference in mid-February 1945, Roosevelt's blood pressure at times approached 250 over 150. The president was examined in March 1945 by cardiologist Howard Bruenn, who reported the president's blood pressure to be 186 over 108 and who diagnosed the president as having congestive heart failure and renal insufficiency. The president's long-term elevated blood pressure was finally having dangerous, harmful effects on his cardiovascular health.

Bruenn prescribed digitalis for the cardiac insufficiency, a low-salt diet, a reduction in the president's substantial alcohol and cigarette use, and bed rest. These were the best tools at Bruenn's disposal. At the time no medicines were available to lower blood pressure.

In April 1945 while posing for a portrait, the president complained of a terrible headache. He then lost consciousness. Bruenn took the president's blood pressure one last time. It was 300 over 190. Two hours later the president was dead. He had suffered a brain hemorrhage, a stroke.

But in 1945 it was not certain what had caused the president's stroke. Was the stroke a consequence of the president's long-term high blood pressure, or was his high blood pressure unrelated to his cardiovascular disease and later stroke? Determining causality is one of the toughest challenges in biomedical science. The answer to this question would not be known with certainty for more than two decades.

The Framingham Heart Study was commissioned by Congress in 1948. A cohort of 5,209 men and women, aged thirty to sixty-two and from the small town of Framingham in eastern

Massachusetts, was selected for a long-term study to investigate the causes of heart disease. The participants had no prior history of heart attack or stroke. The study would be carried out for twenty years, during which time it was expected that many of these participants would develop cardiovascular disease. The study would then determine if there were any aspects of the lives of the participants that were correlated with a higher or lower risk of heart disease or stroke.

When the Framingham Heart Study was initiated, it was generally believed that clogging of the arteries (atherosclerosis) and narrowing of arteries (arteriosclerosis) were normal parts of aging. These age-related changes were considered to be natural and very common if not universal. Similarly, high blood pressure (hypertension) and elevated serum cholesterol (hypercholesterolemia) were also seen as normal and universal consequences of aging. No treatments were available to lower blood pressure or serum cholesterol and it was generally felt that such treatments were not likely to be needed or helpful.

During the next twenty years, the Framingham Heart Study found that cigarette smoking and obesity were associated with a higher risk of heart disease and that exercise was associated with a decreased risk of heart disease. The study also showed that elevated serum cholesterol levels and elevated blood pressure were associated with an increased risk of heart disease.

A number of risk factors for heart disease having been identified, the next question was how to mitigate these risks. Strategies to address the first three risks were pretty straightforward. Patients who smoked were advised to stop smoking. Patients who were overweight were advised to lose weight. Patients who were sedentary were advised to exercise.

But treating patients with high blood pressure or high serum cholesterol was a lot more daunting. You couldn't simply tell people to lower their blood pressure or serum cholesterol. Drugs were needed, and there were essentially no antihypertensive or cholesterol-lowering drugs. The prior thinking, that high blood pressure was normal, had created a disincentive to discover such drugs.

Antihypertensive Drugs

In the early 1950s pharmaceutical scientists pursued the discovery of drugs to inhibit the enzyme carbonic anhydrase. Carbonic anhydrase catalyzes the formation of carbonic acid from carbon dioxide and water. Carbonic acid increases the acidity of the blood and produces effects on the kidneys. The carbonic anhydrase drug that pharmaceutical scientists found, called acetazolamide, acts as a diuretic though its action on the kidneys: it increases the production of urine. As water is lost through urination, blood volume decreases with a concomitant decrease in blood pressure.

In the late 1950s research programs to discover improved versions of acetazolamide led to the discovery of chlorothiazide and a closely related drug, hydrochlorothiazide. These drugs were much better diuretics than acetazolamide and did a better job of lowering blood pressure. But, surprisingly, they were much poorer inhibitors of carbonic anhydrase. It was later shown that the scientists who discovered these drugs had by serendipity discovered drugs that worked via an unanticipated new and superior mode of action.

The thiazides were the best antihypertensive drugs discovered up to that point. They are still very important medications and are used today for patients with mild hypertension. But thiazides are not effective for the treatment of severe hypertension. Unfortunately, the patients with the highest blood pressure, like President Roosevelt, are the ones most in need of blood pressure drugs, and the thiazides were ineffective for such patients.

Epinephrine, which is called adrenaline by British pharmacologists and by the general public in the United States, is a hormone that is best known for its role in the fight-or-flight response. Adrenaline is released into the blood when people encounter a dangerous situation. The released adrenaline prepares you to either fight or flee from danger via four mechanisms. Adrenaline increases blood flow to the muscles, increases heart output, increases blood sugar levels to provide additional energy, and dilates the pupils to allow additional light into the eyes to improve vision.

Adrenaline increases blood pressure, so it would be expected that a drug that blocks its action could potentially be a great treatment for high blood pressure. The problem is that adrenaline produces both excitatory and inhibitory effects on the cardiovascular system, some of which would be detrimental to a patient with high blood pressure. In 1948 the American pharmacologist Raymond P. Ahlquist showed that epinephrine acts on two different types of receptors. His work predicted that a drug that selectively inhibited one of these receptors, the receptor which promotes cardiac activity and which he called the beta-adrenergic receptor, would be expected to lower blood pressure.

In the 1960s Scottish physician and pharmacologist Sir James Black discovered propranolol, the first beta-receptor blocking compound. Propranolol is highly effective in lowering blood pressure, even in patients whose blood pressure is extremely high. Black was awarded the Nobel Prize in Physiology or Medicine for this work in 1988. It seemed that the high blood pressure problem had been solved.

But propranolol is an imperfect drug. In addition to lowering blood pressure, propranolol produces an array of disagreeable side effects in many patients. These include nausea, abdominal pain, and constipation. The drug slows the heart rate in patients with heart failure and worsens the symptoms of asthma. People with high blood pressure feel perfectly fine. Feeling healthy, these patients aren't willing to continue taking a drug that produces unpleasant side effects. Something was needed that worked as well as propranolol but without the side effects.

In the late 1960s two scientists at the Squibb Institute for Medical Research, Dave Cushman and Miguel Ondetti, started a research project to discover an improved drug for the treatment of high blood pressure. They decided to focus their efforts on what at the time was a poorly characterized system for blood pressure control: the renin-angiotensin system. Most researchers in the 1960s thought that the renin-angiotensin system was of only minor importance in the regulation of blood pressure. The general belief was that drugs acting on this system would have small and likely not clinically useful effects on blood pressure control.

But Cushman and Ondetti felt otherwise. They brought in a British scientist named John Vane as a consultant on the project. Vane had recently discovered the mechanism by which aspirin worked, a problem that had frustrated pharmacologists for seventy-five years, and he would go on to win the Nobel Prize for this discovery in 1981. Vane promoted the idea of targeting a particular component of the renin-angiotensin system called angiotensin converting enzyme, or ACE for short. ACE converts a pro-hormone called angiotensin I into an active hormone called angiotensin II, which in turn raises blood pressure. But awkwardly, the only way to move forward with confidence that ACE was a good drug target was to actually discover an ACE inhibitor. In other words, the project would have to be completed in order to know whether or not it might work, a big risk for a hugely expensive undertaking.

Vane proposed a shortcut. He knew of unpublished results demonstrating that the venom from the Brazilian pit viper, one of the world's deadliest snakes, strongly inhibited the renin-angiotensin system. Although the venom could not be turned into a blood pressure lowering drug, it could quickly show whether or not drugs acting on the ACE target would have a substantial effect on lowering blood pressure and thus be worth pursuing.

The Squibb team soon isolated teprotide, the active ingredient in the snake venom, and showed that it did indeed reduce blood pressure, and potently, thus proving that ACE inhibition was a good target for antihypertensive therapy. Although teprotide itself could not be used as a drug, Cushman and Ondetti had proven that their project could produce a valuable drug for the treatment of high blood pressure.

They were ready to move forward, but commercial management at Squibb wasn't supportive and nixed funding for the next stage, the discovery of a clinical candidate compound. Discovering new drugs is an extremely costly and risky endeavor. In the pharma industry only 1–2 percent of drug discovery projects go all the way from conception to regulatory approval and marketing. The rest fail at some point during research programs that take, on average, about fifteen years. If a project is going to fail, everyone hopes it fails quickly. Late failures can cost well over $100 million, with nothing to show for the effort. And even

projects that make it all the way are not always financially successful. A recent analysis by Harvard Business School concluded that
70 percent of approved drugs never earn back their R&D costs.
Thus, it is perhaps not surprising that even though the discovery
of new drugs is essential to support their business, pharmaceutical commercial management is cautious about taking on new
drug discovery projects.

With their antihypertensive research program set to be dis-
continued, Miguel Ondetti was to be reassigned to work on
antibiotics and Dave Cushman to a prostaglandin project. Fortunately, however, two key R&D managers, Zola Horovitz, the
pharmacology department director, and Arnold Welch, the head
of research, were willing to take the risk of backing Cushman and
Ondetti in the absence of commercial support. The antihypertensive project thus continued.

The team's next step was to test two thousand diverse chemical structures for ACE inhibition, but no hint of useful activity was
seen with any of the tested compounds. Arguably, things were not
looking good. But on March 13, 1974, the ACE inhibition team
came across a paper on carboxypeptidase A, an enzyme that was
in many ways similar to ACE. The paper described the rational
design of an inhibitor of carboxypeptidase A based on the catalytic action of the enzyme and the structure of its substrate, the
chemical upon which the enzyme acted. The team realized that,
based on their knowledge of the catalytic action of ACE and the
structure of the substrate, angiotensin I, they could take a similar
rational approach to design an inhibitor of ACE.

A year and a half later the team discovered captopril. This was
lightning-fast work for a small team, and the approach they took
is now hailed as the model for the rational design strategy in drug

discovery. Cushman and Ondetti went on to win the 1999 Lasker Medical Research Award (often called America's Nobel) for this discovery.

But there was a problem. Commercial management was still not interested. Clinical studies are far more expensive than pre-clinical laboratory experiments. Typically, the cost to discover a clinical candidate drug, like captopril, is less than 20 percent of the total R&D expense required to bring the drug to market. The commercial managers didn't want to spend a lot of additional R&D money on what they still considered to be a risky project. After all, many highly respected scientists in the field claimed that drugs inhibiting ACE were unlikely to be useful. The commercial managers were also worried about maintaining their profits from nadolol, Squibb's own propranolol look-alike drug. Nadolol had decent sales, and the managers argued that if patients taking nadolol switched to captopril, the result wouldn't be a net increase in sales. The project was over.

Despite the disinterest on the part of commercial management, Cushman and Ondetti remained proud of their work and wanted to publish their findings for the scientific community. Since patent applications had already been filed, they argued that there was no risk in publishing should commercial management change their minds about commercializing captopril. After a lot of arguing, they were finally allowed to publish their work but with the restriction that they spend their full time on their new projects and prepare the captopril manuscripts only on their own time, at night and on weekends. Over the next few years Cushman and Ondetti completed three papers and published them in highly respected scientific journals.

Soon after the papers appeared, Squibb was contacted by

several top-notch American medical schools expressing interest in participating in the clinical trials for captopril—the very trials that had been nixed! This outside interest induced commercial management to rethink their decision. If, in their view, Cushman and Ondetti might not be objective in promoting their pet project, the scientists at top-tier medical schools had no reason to be biased in supporting captopril. The medical school professors were interested only in helping patients.

Commercial management therefore reversed their decision and approved funding the clinical development of captopril. Although unanticipated side effects were reported in early trials—overconfidence resulted in overdosing the drug—later studies showed that lower and still highly effective doses were safe, well tolerated, and highly effective treatment for hypertension. Captopril was approved by the FDA on April 6, 1981, and soon went on to become a blockbuster, with annual sales of greater than $1 billion.

Captopril was wildly successful financially, earning Squibb more money than even its most optimistic supporters expected. Soon Bristol-Myers realized that Squibb's stock price didn't reflect the financial prospects for captopril. They launched a successful hostile takeover of the company in 1989 that resulted in creation of Bristol-Myers Squibb. Following the $12 billion Bristol-Myers merger offer, the Squibb stock price rocketed from $24.75 to $112.50 per share. Ironically, the commercial success of captopril led to the demise of Squibb as an independent pharmaceutical company.

While overall the project had been highly successful, the lack of good teamwork between R&D and commercial management resulted in significant missed opportunities. Following the commercial launch of captopril, the emotional arguments, intense disputes, and high level of acrimony left both sides mentally and emotionally exhausted. Captopril was not a perfect drug: it had to be dosed two to three times a day and had some side effects, although not nearly as bad as propranolol's. Many patients complained about the frequent dosing and reported that the drug produced an unpleasant metallic taste. (Long-term dosed drugs will be more successful if they are clean of side effects and are easy to take.) Captopril's liabilities were quickly appreciated by Merck, which, after reading the Cushman and Ondetti papers, started their own program to come up with improved versions of captopril, something Squibb could have easily done but chose not to.

Merck quickly designed compounds that had improved dosing and superior tolerability: enalapril (Vasotec) and lisinopril (Prinivil). Lisinopril is now a standard of care for the management of hypertension. Squibb made a lot of money but lost significant market share to these and other competing compounds that were quickly discovered based on the Squibb publications. Squibb could have made a lot more money if they had accelerated the development of captopril while delaying publication of the papers and then launched research efforts to discover improved follow-on compounds. But commercial management's suspicions that their own scientists had lost objectivity delayed supporting the project and gave competitors a great head start to find superior competing drugs.

Dave Cushman and Miguel Ondetti

Dave Cushman was born in Indianapolis, Indiana, in 1939. After finishing high school, Cushman attended Wabash College in Crawfordsville, Indiana, studying zoology and chemistry. He had grown up poor and was the first in his family to attend college, often saying, "Being poor is a great stimulus for wanting to achieve something." After graduating from Wabash College, Cushman earned his PhD from the University of Illinois in 1966 and immediately joined the Squibb Institute for Medical Research, the only place he ever worked.

Cushman retired in 1994 as a distinguished research fellow in the Department of Cardiovascular Biochemistry at Bristol-Myers Squibb Pharmaceutical Research Institute. Tragically, Dave Cushman died from cardiovascular disease in 2000 at the age of sixty, a disease from which his drug had protected millions of patients.

Miguel Ondetti was born in Buenos Aires, Argentina, in 1930 to working-class immigrant parents from Italy. Despite his intense interest in science, Miguel and his brother were sent to a vocational high school, where they studied bookkeeping and accounting. Ondetti was not able to earn a high school baccalaureate at his vocational high school and, as a result, was denied admission to the University of Buenos Aires. He spent an additional two years auditing classes in order to earn his baccalaureate and thereafter obtained admission at the University of Buenos Aires in their chemistry program.

While at the university, Ondetti supported himself by working as a bookkeeper. He took early shifts at the office so that he could attend required laboratory classes in the afternoon. His university program was similar to a combined BA/MS program in American education. Because opportunities for scientists in Argentina were limited, the program provided a broad exposure to chemistry to prepare students for a wide variety of jobs, especially in the paint, petrochemical, and pharmaceutical industries, which were the best job opportunities at the time for chemists in Argentina.

After five years of study, Ondetti earned a licentiate degree in 1955. This was a time of political turmoil and slow economic growth in Argentina. Many university professors had been forced to move to industry jobs, including the University of Buenos Aires head of organic chemistry, Dr. Venancio Deulofeu, who took a position at the Squibb Institute for Medical Research in Buenos Aires. In view of the economic and political situation at the time, Ondetti's best option for further education was to accept a research training scholarship at the Squibb Institute.

Ondetti completed his PhD studies there under the direction of Dr. Deulofeu and in 1957 was awarded a PhD from the University of Buenos Aires. After earning his PhD, he accepted a full-time research position at the Squibb Institute.

In March of 1960 the head of the New Jersey–based Squibb Institute was in Buenos Aires for his annual visit to the foreign subsidiary. He scheduled an interview with Miguel Ondetti, during which he offered him a permanent position with the Squibb Institute in the United States. He gave Ondetti only two days to think about it. A serious problem was that at exactly that time Ondetti's wife was preparing to open a dental office in Buenos Aires. Accepting the job meant that Ondetti's wife would have to forgo a major professional opportunity.

The couple decided to take a chance and never looked back. There were other issues when the Ondettis arrived in New Jersey. Miguel Ondetti had learned English from chemistry textbooks, which didn't exactly provide the vocabulary needed for everyday activities of life in New Jersey. Plus, he had been taught British English, which is different than the English spoken in the United States. Ondetti referred to things using terms unknown in the US: spanner, hob, car wing, etc. But eventually things worked out. The

Ondettis remained in New Jersey for forty-four years and raised two children there.

Miguel Ondetti retired from Bristol-Myers Squibb in 1995 and for many years afterwards had a productive consulting practice, advising numerous companies on their drug discovery research projects. During these years, I remember seeing Miguel at scientific conferences, where he was very much in demand.

In addition to sharing the 1999 Albert Lasker Award for Clinical Medical Research, Cushman and Ondetti shared the 1982 American Chemical Society GlaxoSmithKline Alfred Burger Award in Medicinal Chemistry and were named Heroes of Chemistry by the American Chemical Society. Miguel Ondetti was also awarded the 1991 Perkin Medal by the American Chemical Society, their highest recognition for achievement in applied chemistry. Ondetti died in August 2004 at the age of seventy-four.

Dave and Miguel were very much the odd couple of pharmaceutical research. Dave Cushman was lighthearted, always cheerful, and never ever seemed serious. He always had a mischievous look on his face, as if he were preparing to get into some minor trouble or to play a trick someone. The first time I saw Dave, he was at the departmental copy machine running off copies of a comic book that he wanted to share with the entire department. I asked myself, what kind of adult would waste their time copying comic books? Dave was a truly passionate person—about science and what he considered to be a great comic book.

In contrast, Miguel Ondetti was an extraordinarily serious person. When asked a question, he would briefly pause with a distant look in his eyes, while he thought through the perfect response. Although most of the questions he fielded were scientific, he acted the same way in response to almost any question. If you saw Miguel in the cafeteria and asked him what he was planning on having for lunch, he would pause and think it about for a little while to make sure that he was able to accurately and completely tell you what he was about to order: turkey on rye with extra mayonnaise and a half sour pickle on the side, or something like that.

This odd couple made one of the most important pharmaceutical discoveries of the twentieth century, against long odds and by battling against both the scientific and commercial establishments. Their simple strategy was to publish, disseminating their work to the research community and then to let good science speak for itself. And that is exactly what happened.

Cushman and Ondetti's story illustrates a different sort of hurdle in getting one's work accepted. Basic science is vetted by the scientific establishment, and acceptance comes from that community. But applied science is appraised according to both scientific and financial criteria, and the two systems often operate in opposing ways. Great science may not have immediate or even long-term commercial value, while bad science can be extremely profitable—consider the many beauty aids and nutritional supplements that basically do not work yet make their purveyors a lot of money anyway.

12 | Surendra Nath "Suren" Sehgal
Discovered a major drug for organ transplant patients

Organ Transplantation

Ancient people imagined that it could be possible to reattach severed body parts (a process that today we call autotransplantation) or to replace lost body parts with spares harvested from animals (xenograft transplantation) or other people, individuals unrelated to the host (allograft transplantation). Ancient Egyptians, Greeks, and Romans all described transplantations of bone, skin, and teeth. The Christian Bible includes multiple examples of autotransplantation. Jesus of Nazareth restored a servant's ear that had been severed in battle by Simon Peter's sword. Saint Peter reimplanted the breasts of Saint Agatha after she was tortured and mutilated. Saint Mark reimplanted the battle-amputated hand of a soldier.

An account of allograph transplantation appears in the thirteenth-century book *The Golden Legend* (*Legenda Aurea*), a collection of stories about Christian saints and festivals written by Jacobus de Voragine. In *The Golden Legend* de Voragine tells the "miracle of the black leg," a story in which the gangrenous leg of the Roman deacon Justinian was amputated and replaced with the leg of a dead Ethiopian man.

All these stories are fictional and aspirational. But transplantation seemed to ancient peoples to be a very reasonable strategy, if only the right replacement technique could be developed. It seemed that everything would work out fine if you could just figure out how to successfully make the attachment or reattachment.

Scientific studies to develop surgical methods for organ transplantation were first pursued in the late nineteenth century. One of the first attempts was related to the surgical treatment of goiter, the enlargement of the thyroid gland in the neck. Not knowing that the thyroid gland is an essential organ, surgeons developed procedures to remove it. One early master of thyroidectomy surgery was the Swiss surgeon Theodor Kocher (1841–1917), who successfully removed entire thyroid glands from his patients without immediate adverse effects.

But then years after performing the procedure, his patients developed a syndrome that we now know as hypothyroidism, called cretinism in children. The thyroid gland produces thyroid hormones, which serve as master regulators of metabolism. People lacking adequate thyroid hormone develop a wide variety of symptoms: poor appetite, cold intolerance, constipation, lethargy, and weight gain.

Thyroid hormones weren't available to treat thyroidectomy patients at the time. In 1891 a British physician named George Murray described a method for treating hypothyroidism patients with animal thyroid extracts, but results were inconsistent and the approach was not accepted by the medical community. We now know that most often the inconsistency stemmed from simple rookie procedural errors. For example, one group of therapeutic failures were later found to be due to the fact the butcher responsible for supplying animal thyroid gland mistakenly supplied the physician with the thymus gland instead of thyroid. Success with this treatment was finally achieved after thyroxin was first purified from the thyroid gland in 1916.

Without thyroid treatments, Kocher's only therapeutic alternative for his patients who had become ill following the removal of their thyroid gland was to try to reverse the procedure and transplant a thyroid gland from a very recently deceased unrelated individual. But Kocher's transplantation surgery did not produce long-term cures. The transplanted thyroid gland was soon rejected by the patient for reasons that were unknown at the time. It was thought that the likely problem was the lack of adequate surgical techniques needed to solve the pure "plumbing" problems of transplantation: attachment or reattachment. These pure surgical challenges were soon overcome.

During the first decade of the twentieth century, the French-American surgeon Alexis Carrel designed the arterial clamp, a device that made it possible to temporarily interrupt blood flow through a vessel, thus enabling surgeons to perform the fine surgical dissection needed for organ transplant. Carrel also developed a blood vessel suturing technique that made it possible to permanently connect blood vessels in organ transplantation procedures. Carrel was awarded the 1912 Nobel Prize in Physiology or Medicine for his work.

But Carrel soon became aware that there was still another problem that needed to be overcome: graft rejection by a foreign host. Graft rejection had been commonly observed by surgeons but was little appreciated or understood at the time. In 1910 Carrel observed insightfully: "Should an organ, extirpated from an animal and replanted into its owner by a certain technique, continue to functionate normally, and should it cease to functionate normally when transplanted into another animal by the same technique, the physiological disturbance could not be considered as brought about by surgical factors. The changes undergone by the organ would be due to the influence of the host, that is, to biological factors."

That is, even once the surgical issues had been solved, there appeared now to be a new barrier, the acceptance of the graft by the host, a barrier that would have to be overcome for successful organ transplantation. It would take many decades of research for that problem to be fully described and understood.

Graft Rejection

There existed two possibilities to explain why organ transplantation grafts were rejected. The first was that there was some unknown inadequacy in the established surgical techniques; something that needed to be done and was not being done or that was needed to be done differently: things that no surgeon was likely to be remotely aware of or could even imagine. The second possibility was that the surgical techniques were fine but there was an unrelated and unknown process that blocked the transplant from working. The question was finally definitively answered in 1954.

On December 23, 1954, a surgical team led by Joseph Murray at the Brigham Hospital in Boston removed a kidney from a healthy donor and transplanted it into his identical twin, who was dying of renal disease. The operation was successful and the organ began to function immediately. More importantly, the transplant recipient survived for nine years and, when he eventually died, it was from causes unrelated to the kidney transplantation surgery. The donor survived for fifty years.

In 1990 Joseph Murray shared the Nobel Prize with E. Donnall Thomas, a physician scientist who had in his own work shown that successful transplantation required a match between the transplanted organ and the immune system bone marrow cells in the host. The prize was awarded for "for their discoveries concerning organ and cell transplantation in the treatment of human disease."

This first successful transplantation was performed against a very long background of failure. Medical professionals were becoming despondent about the possibility of ever developing an effective organ transplantation procedure, and here was a true success story, although one based on a singular unique circumstance. It proved that the surgical techniques that had been developed were effective. It was now necessary to refocus research on the understanding the role of the immune system in organ transplantation rejection.

Peter Medawar and Acquired Immune Tolerance

The British scientist Peter Medawar earned an honors degree in zoology from Magdalen College, Oxford, in 1935 followed by a doctoral degree from the Sir William Dunn School of Pathology, Oxford, in immunology in 1941. During World War II Medawar worked on treatments for the skin wounds of injured soldiers. In 1951 he developed a new method for skin grafting in mammals, a method that enabled him to discover acquired immune tolerance. In 1949 an Australian scientist named Frank Macfarlane Burnet had published a hypothesis stating that during embryonic life cells gradually acquire the ability to distinguish between their own cells and intruding foreign cells. If true, the ability to distinguish between self and foreign cells was acquired and not innate, and it might be possible to manipulate self-recognition for organ transplantation. Medawar set out to use his new skin grafting method to test Burnet's idea.

Medawar and his team extracted cells from young mouse embryos and injected them into mouse embryos of different strains. When these injected mice embryos developed into adults, they were then challenged by grafting skin onto them from the cell donor strain. Mice that had not received any extracted cells rejected the grafts. But mice that had received the extracted cells showed no graft rejection. The injection of foreign cells into the mouse embryos had allowed the embryos as adults to accept the foreign cells. These injected mice had acquired immune tolerance. Medawar had thus shown that it was possible to manipulate a host to make it tolerate foreign tissue.

Medawar is regarded as the "father of transplantation." He and Burnet shared the 1960 Nobel Prize in Physiology or Medicine "for discovery of acquired immunological tolerance." Sadly, just ten years after receiving the Nobel Prize, Medawar suffered a severe stroke that impaired both his speech and movement. He thereafter was able to continue doing only a limited amount of scientific research and only with considerable physical support from his wife.

In the late 1970s Medawar was invited to lecture at Princeton in the Gauss Seminars in Criticism. The Gauss Seminars were designed to provide a focus for discussion, study, and the exchange of ideas in the humanities and to be conducted by guests invited to present material on which they were working. Over the years Gauss seminar leaders included Erich Auerbach, Hannah Arendt, W. H. Auden, Noam Chomsky, Roman Jakobson, Elaine Scarry, Joan Scott, and Raymond Bellour. Faculty and graduate students from Princeton University, the Princeton Institute for Advanced Study, and the community at large were encouraged to participate in each seminar.

As a biological science graduate student, I generally took little interest in the Gauss seminars. They were for liberal arts humanities students. But I jumped at the opportunity to attend when I learned that Peter Medawar had been invited to lecture. I remember Medawar, with his wife by his side, speaking on the history and philosophy of science. Despite his obvious physical impairments, his talks were scintillating. In particular I remember his delightful discussion of how Galileo's observations and logical analyses had made it possible for him to prove that the earth revolved around the sun rather than vice versa. But after his lectures, Medawar apologized to the audience, saying that his impairments had prevented him from pursuing the high level of science in which he had been able to engage prior to his stroke. What he had lectured on seemed pretty high level to me and a true tribute to Medawar's peerless intellect, although also a disturbing commentary on my own standards for excellence at the time.

While Medawar had proven that it was possible to manipulate animals to produce immune tolerance, he did not design a clinically applicable immune tolerance method for transplant patients. Based upon Medawar's findings, however, physician scientists began to develop and test approaches to produce immune tolerance in the clinic. The first of these was total-body irradiation. This procedure destroyed the immune system and prevented rejection. But it also quickly led to the death of the patient from overwhelming infections. Total-body irradiation was not the solution.

Investigators then turned to cancer chemotherapy drugs, which were known to produce extreme immunosuppression in addition to killing cancer cells. One of the first of these drugs was azathioprine, which was shown to prolong kidney transplants in dogs. Corticosteroids (drugs like hydrocortisone) also suppress the immune system. Soon drug cocktails containing azathioprine and a corticosteroid were being used to try to increase graft survival in transplant patients.

While the azathioprine-plus-corticosteroid cocktails worked, the mortality of renal transplantation patients was still high, about 40 percent at one year post surgery. Better drugs were urgently needed to treat transplant patients.

Surendra Nath "Suren" Sehgal and Rapamycin

Suren Sehgal was born in 1932 in Khushab, a small Indian village that is now part of Pakistan. His father owned a pharmaceutical factory and encouraged Suren to study pharmaceutical science. Sehgal earned B. Pharm and M. Pharm degrees from Banaras Hindu University in 1952 and 1953 and then earned a PhD from Bristol University in England in 1957. He went on to pursue postdoctoral studies at the National Research Council of Canada.

After completing his postdoctoral studies in 1959, Sehgal was recruited by Roger Gaudry, the director of research at Ayerst Pharmaceuticals in Montreal, Canada, to join the microbiology research team there. At Ayerst he went on to recruit a number of his local scientist friends to work with him, forming the core of the microbiology research department at Ayerst.

In 1964 a Canadian scientific expedition traveled to Easter Island to gather exotic plant and soil samples, which they shared with Ayerst's microbiology department. Screening the soil samples, the department discovered a new antifungal antibiotic produced by a soil microbe. Ayerst named the drug rapamycin after the Polynesian name for Easter Island, Rapa Nui. Rapamycin is a highly potent antifungal drug, but further testing showed that the compound also powerfully suppressed the immune system. Antibiotics need to work in concert with the immune system in order to cure infections, so rapamycin was not going to be a clinically useful antifungal agent.

It is a common misapprehension that new drugs are designed by an engineering process, like computers, mobile phones, or electric cars. But this is not true. In reality, drugs are discovered, and as a result you find what you find, which might not be what you were initially searching for. The challenge then becomes to see whether you can come up with a clinical use for whatever you found, plus a way to make it economically viable. Sehgal had set out to find an antifungal drug but instead found an immunosuppressive drug.

Sehgal soon became intrigued by the drug's immunosuppressive properties and suspected it might prove useful as a treatment for organ transplant patients. Unfortunately, Ayerst didn't have an interest in drugs for organ transplants, so commercial management decided to terminate the project. Sehgal protested. He repeatedly tried to convince management to develop his drug. Research management soon became annoyed with his persistent entreaties and came up with a strategy to shut him up for good. Sehgal was ordered to kill the rapamycin-producing soil microbe via autoclave sterilization.

Once the soil microbe that made rapamycin was dead, there would be no way to make the compound and thus no possible further argument about pursuing the project. Although Sehgal was an extremely honest man, this time he did something a little devious. He sterilized the culture as ordered, but also secretly brought a replicate culture home with him that he stored in his basement freezer. His hope was that as management changed, one day a new manager would be hired who would support the development of his immunosuppressive drug.

There is a humorous anecdote that periodically circulates among scientists in the pharmaceutical industry. The head of research at a major pharmaceutical company is sacked and a replacement is hired. The day the new head of research arrives at the company he is greeted by his outgoing predecessor, who hands him two sealed envelopes. The predecessor tells the new head of research, "If after the first six months things start going badly, open the first envelope. And after two more years if things continue to go badly, open the second envelope."

Although the new director starts out with big plans and high expectations, the implementation of his new strategy is thwarted by a number of unexpected problems and simple bad luck. The CEO expresses concern. So after six months the new director opens the first envelope. There is a single sheet of paper inside on which is written: "Blame all problems on the prior director."

The new director goes to the CEO and complains that the prior director had left him with a chaotic mess. He tells the CEO that when he started the job the whole R&D operation had been suffering from incompetent staffing. Many of the incumbent scientists were clearly not up to the job and had to be replaced. It had taken him six months just to right the ship. The CEO is totally sympathetic. He tells the new director not to worry and just to keep up his hard work to discover new drugs.

But two years later things are still not going well. One drug candidate had to be dropped over toxicity concerns. Another produced severe unanticipated side effects. A third major project had to be stopped over the issuance of a blocking patent to a competitor. Commercial management is highly dissatisfied. The new director therefore opens the second envelope. There is a single sheet of paper inside on which is written: "Prepare two envelopes." Research management turns over frequently.

During Sehgal's employment, Ayerst Pharmaceuticals was owned by American Home Products (AHP), a holding company first established in 1926. The holding company philosophy was central to AHP's business strategy and compounded management resistance to Sehgal's efforts to develop rapamycin as an organ transplantation therapy. AHP had no allegiance to any particular industry, product group, or business sector. They bought and sold consumer goods companies based solely upon their financial performance. In the last half of the twentieth century AHP products included Dristan nasal spray, Anacin headache remedy, Preparation H for hemorrhoids, Chef Boyardee spaghetti, Mama Leone's pasta, Gulden's mustard, Wrigley's chewing gum, Woolite laundry detergent, Black Flag insecticide, Old English furniture polish, PAM cooking spray, and a line of aluminum pots and pans. AHP also owned two pharmaceutical companies, Canadian Ayerst Pharmaceuticals and Wyeth Pharmaceuticals, an old-line American pharmaceutical company.

Long-term R&D investments do not pay off when subsidiaries are bought and sold for short-term gain. Some AHP managers argued against supporting any R&D at all, saying that if they needed new products to support their businesses, they could negotiate good terms to buy them from other companies. AHP commercial management knew what they wanted, and were annoyed when insubordinate scientists in the R&D organization promoted something outside their carefully prepared business plans.

AHP was so invested in their holding company philosophy that they didn't merge their two pharmaceutical companies until 1987, waiting almost forty-five years despite the obvious economies-of-scale cost savings that would result. It is said that the Wyeth-Ayerst merger was one of the most contentious in the pharmaceutical industry. For four and a half decades the two companies were operated as bitter rivals competing for corporate resources and had built up a long history of ill will and animosity. Fortunately for Sehgal, in addition to uniting the two companies, the merger brought about major changes in management and a change in management philosophy.

A novel immunosuppressant therapy called cyclosporine (Sandimmune) had been discovered by Sandoz Pharmaceuticals in 1971 and brought into medical use in 1983. Cyclosporin was an entirely new type of drug that suppressed the immune system in a completely new way. Cyclosporin depressed the activity of T-cells in the immune system via inhibiting an enzyme called calcineurin. Calcineurin is a key regulator of the immune system T-cells that are a major determinant of transplant rejection. Its inhibition blocks T-cell activation. It is thus a totally different drug from azathioprine, which basically acts by killing all rapidly dividing cells, including activated immune cells. Cyclosporin is much more selective than azathioprine in its action. It can be safely taken at doses that strongly disable the immune system without producing the potent toxic effects seen azathioprine therapy. After its clinical introduction, cyclosporin increased the rate of one-year graft survival to 80 percent.

But no one at Ayerst was very impressed with the cyclosporin discovery, nor did they think that a similar drug would be a profitable venture for them. Fortunately for Sehgal and AHP, a drug very similar to rapamycin called tacrolimus was discovered by Fujisawa Pharmaceutical Company in 1987. Fujisawa decided to advance the drug into clinical development as an immunosuppressant therapeutic. The merger of Wyeth and Ayerst along with the concomitant management changes occurred at exactly this time. The new managers of Wyeth and Ayerst felt that if Fujisawa had found a financial opportunity with their drug, it might be worthwhile for AHP to pursue something similar.

So Sehgal finally found support for his project, two decades after he started it. Sehgal went home and fished the package containing the rapamycin-producing microbe from his basement freezer—a package he had carefully labeled DO NOT EAT, protecting it for many years from his hungry family.

Rapamycin, now renamed sirolimus, was finally approved in 1999 for use as an immunosuppressive drug in renal transplantation. Sirolimus is also used as a coating for coronary stents to prevent restenosis and more recently has been approved for the treatment of lymphangioleiomyomatosis, a rare progressive neoplastic disease producing thin-walled cysts in the lungs and angiomyolipomas in the kidneys. Commercial manufacturing of sirolimus is based upon the fermentation of the microbe that had lain undisturbed for decades in Sehgal's basement freezer.

Beyond its clinical use, studies of the sirolimus mechanism of action have led to the discovery of a major new cell regulatory pathway, a pathway that connects cellular metabolism and growth. Rapamycin inhibits a protein called mTOR, a global regulator of cell growth, cell proliferation, cell motility, cell survival, and protein synthesis. While rapamycin does not inhibit the same target as cyclosporin, both drugs selectively suppress the immune system. In the immune system mTOR is a key sensor of a wide range of immune inputs and, by so doing, regulates the ability of immune cells to respond to foreign antigens like those produced by a grafted organ. When mTOR is inhibited, the immune system loses the ability to respond to a foreign organ.

In 1998 Suren was diagnosed with stage 4 metastatic colon cancer following a routine colonoscopy. He completed his laboratory studies with rapamycin and retired after forty years of service. Suren continued his rapamycin work to the very end of his life as a consultant to Ayerst (now renamed Wyeth), completing work on his last scientific publication just a few weeks before his death. During his own cancer treatment, Sehgal met numerous kidney transplant patients who were surviving thanks to his drug. He underwent major liver surgery and tried to extend his own life by experimentally treating himself with rapamycin to stave off the cancer metastases that had spread to his liver. But his cancer was unstoppable, even with rapamycin. Sehgal died in January 2003, leaving behind his wife, children, many friends and colleagues, and many thankful transplantation patients.

Sehgal faced some of the same challenges as Cushman and Ondetti: battling to achieve both scientific and financial recognition for his invention. But Sehgal employed a strategy different from the one employed by Cushman and Ondetti. Cushman and Ondetti presented their results to the scientific community, hoping that famous scientists would convince Squibb commercial management to pursue their drug. In contrast, Sehgal realized that commercial decisions are often subjective. Different managers make different decisions. So he just bided his time until a new manager appeared who would support his project.

13 | Alfred Wegener
Formulated the theory of continental drift

The biology department, when I was a graduate student in biology at Princeton University during the 1970s, was located in Guyot Hall, an academic building shared with the geology department. The biologists were on the east side of the building, and the geologists on the west side. We also shared a parking lot. One of the geology graduate students had a car with an odd REUNITE GONDWANALAND bumper sticker. This was during the Vietnam War, and lots of students had political bumper stickers on their cars. But "Reunite Gondwanaland" seemed to have nothing to do with any of the political issues of the time.

One day I came across a scientific article that revealed what this bumper sticker was about. It was both a humorous commentary on contemporary politics (think "Reunite South and North Vietnam") and a reference to a geological theory that had been proposed during the early twentieth century but rejected by the geology establishment for more than fifty years. One of the professors in the geology department had recently published major new findings that finally led to the scientific acceptance of this half-century-old idea.

Alfred Wegener

Alfred Wegener was born in Berlin in 1880. He attended the Köllnisches Gymnasium on Wallstrasse in the eastern part of Berlin (named after long-ago demolished eighteenth-century fortifications, not the twentieth-century Berlin Wall), graduating as the best in his class. Wegener was interested in science, but he never focused on a single scientific discipline. After high school he studied physics, meteorology, and astronomy at universities in Berlin, Heidelberg, and Innsbruck. Wegener obtained a doctorate in astronomy in 1905 from Friedrich Wilhelms University (today Humboldt University) in Berlin. But in later life his research studies focused instead on the fields of meteorology and climatology.

Wegener's first academic appointment was as a lecturer in meteorology, applied astronomy, and cosmic physics at the University of Marburg. In 1912 Wegener presented his theory of continental drift, initially as a lecture to the Geologischen Vereinigung at the Senckenberg Museum, Frankfurt am Main. Later that year he published his ideas in a long three-part article, "Die Entstehung der Kontinente" (The Formation of the Continents) in the journal *Dr. A. Petermanns Mitteilungen aus Justus Perthes' geographischer Anstalt* and then in a shorter summary form in the journal *Geologische Rundschau*. Almost immediately, his theory was harshly rejected by the geology scientific community.

Wegener's scientific work was interrupted by World War I, in which he served as a German infantry officer on the Belgian front. After being twice wounded, Wegener was transferred to the army weather service for the duration of the war. When it was over, no German university would give him a tenure-track faculty appointment, the result of the strong negative reactions to his controversial continental drift theory.

The best opportunity he could find was a nonacademic position as a meteorologist at the German Naval Observatory (Deutsche Seewarte) in Hamburg. In 1921 he was appointed to a temporary senior lecturer position at the new University of Hamburg. Finally, in 1924 Wegener obtained a professorship in meteorology and geophysics at the University of Graz in Austria. He at last had a secure tenured position, although not in Germany. At Graz he studied physics, the optics of the atmosphere, and the properties of tornadoes.

The first decades of the twentieth century were the concluding years of the Age of Exploration, now focused on expeditions to the North and South Poles to complete the cartographic description of the world. During his career, Wegener, caught up in the exciting enterprise of putting the final touches on the world map, participated in or led four extremely dangerous expeditions to Greenland in order to study the Arctic climate.

On the first expedition, in 1906, two of his colleagues died on an exploratory trip by dogsled. During the second expedition in 1912–1913, the expedition leader, Johan Peter Koch, broke his leg when he fell into a glacier crevasse and had to spend months recovering from his injuries. Despite this setback, Wegener and Koch were the first expedition team to winter on the inland ice in northeast Greenland. In the summer of 1913, four members of the second expedition crossed the inland ice but struggled to make their way across difficult glacial breakup terrain and ran out of food only a few kilometers from the western Greenland settlement of Kangersuatsiaq. At the last moment, and after the final pony and dog had been eaten, the team was discovered by a Greenlander who by good luck just happened to be in the right place at the right time.

In 1929 Wegener made a third trip to Greenland to test equipment for a fourth expedition. The fourth expedition, funded by the German government, was launched in 1930 with a team of fourteen participants. This expedition planned to set up a two-man overwinter camp in the Greenland mid-ice (at a site called Eismitte). But a late thaw delayed provisioning the mid-ice camp by six weeks, which resulted in inadequate supplies being laid in. In the early fall the men in the mid-ice camp sent a message to the base camp, saying that they had insufficient fuel stores and would have to return to west base camp on October 20.

Unwilling to abandon the mid-ice camp and motivated by a desire to show the German government he was making good use of their financial support, Wegener set out with a team in late September via dogsled to bring more supplies the camp. Temperatures in late September reached –60°C (–76°F). Wegener's meteorologist on the resupply team developed extreme frostbite, and his toes had to be amputated with a pocketknife, without the benefit of anesthesia. Twelve of the fifteen men who started the resupply mission turned back. The remaining three arrived at the mid-ice camp on October 19.

There were enough supplies for only one additional man to overwinter there, so Wegener and another team member, Rasmus Villumsen, decided to return to their west base camp. They never made it. The following spring a rescue mission discovered Wegener's body. Villumsen's remains were never found. Wegener was fifty years old when he died.

During his life, Wegener made many contributions in a wide variety of scientific disciplines. We now know that the most important of these was the theory of continental drift that he had published in 1912, eighteen years prior to his death. But his theory was roundly rejected by the scientific establishment, and Wegener didn't live long enough to convincingly promote it or to see its eventual acceptance by his peers.

The Age of Exploration

The Age of Exploration was a period lasting from approximately the fifteenth century to the early twentieth century, when sea-faring Europeans explored regions around the globe. Exotic places like China had long been well known to the Europeans, but only through long-established traditional land routes, like the Silk Road. Major parts of the world, accessible only by sea, were unknown and virtually undreamt of.

Naval explorations, initially led by the Portuguese, eventually delineated the entire map of the Earth. These journeys started with expeditions by Portugal to the Canary Islands. The discovery of the Canary Islands was quickly followed by the discoveries of the Madeira and Azores archipelagos. The coast of West Africa was then explored, leading to the establishment of a sea route to India in 1498 by the Portuguese mariner Vasco da Gama.

Spain sponsored transatlantic voyages by Christopher Columbus, leading to the European discovery of the Americas and later to the first circumnavigation of the globe by Ferdinand Magellan. While Spain and Portugal concentrated on South America, later naval expeditions conducted by the Russians, French, Dutch, and English produced detailed maps of North America, Asia, Australia, and New Zealand as well as more complete maps of the Atlantic, Indian, and Pacific Oceans. Final explorations of the polar regions, carried out during the early twentieth century, completed the geographical description of the earth.

Continental Drift

Abraham Ortelius was a sixteenth-century Dutch mapmaker who created the first modern atlas, the *Theatrum Orbis Terrarum* (*Theater of the World*), in 1570. Ortelius was the first person to point out the remarkable geometrical similarity between the coasts of the Americas and Europe-Africa, and in particular the west coast of Africa and the east coast of South America. In his book *Thesaurus Geographicus*, Ortelius wrote that it seemed like the Americas had been torn away from Europe and Africa.

For a period of almost 350 years, many other scholars continued to write about this symmetry and to speculate on what it implied about the process that produced the continents on the world map. Despite the fact that he was a meteorologist and a climatologist and thus hardly an expert in the field of geology, Alfred Wegener was part of this tradition. When he publicly presented his ideas on the shapes of the continents and their transformations in January 1912, Wegener proposed that the seven current modern continents had once been a single landmass that he called the *Urkontinent* (Pangaea in English).

According to his theory, the original landmass broke apart, and the broken pieces later moved to new locations to produce the current world map. We now know that Pangea had been composed of two giant subcontinents, Laurasia and Gondwanaland. Breaking into smaller pieces, Laurasia produced the northern land masses, North America, Greenland, Europe, and northern Asia. Gondwanaland produced the southern land masses, South America, Africa, India, Madagascar, Australia, and Antarctica. In his 1912 presentation, and then in two subsequent publications, Wegener called the movement of these pieces of Pangaea to their current locations continental drift (in German *die Verschiebung der Kontinente*).

In support of his theory, Wegener brought together a wide variety of observations. First of all, there was the cartological evidence, that the shapes of the continents were such that they could easily fit together like a giant jigsaw puzzle to produce a single combined landmass structure. He argued that the fit was too precise to have been an accident, and he adduced other evidence in support of his theory.

Second, there are two examples of similarities in rock formations on different continents on opposite sides of the Atlantic Ocean. In the Northern Hemisphere the Appalachian Mountains in the United States and Caledonian Mountains in Scotland share common features. Both ranges reveal elongated belts of folded and thrust-faulted marine sedimentary rocks, volcanic rocks, and slivers of ancient ocean floor, evidence that these rocks, now divided by an ocean, were formed together. In the Southern Hemisphere, the Karoo strata in South Africa and Santa Catarina rocks in Brazil strongly resemble one another. In these mountain systems, the rock strata share identical layers of sandstone, shale, and clay laced with seams of coal. Also, the mountain ranges on opposite sides of the Atlantic Ocean in both the Northern and Southern Hemispheres show similar weathering, suggesting that they were all originally part of a single entity (i.e., Pangaea) very long ago. Current scholarship indicates that this weathering occurred during the Permian period (300 to 250 million years ago) and, for some of the mountains, additional weathering occurred during the Jurassic period (200 to 150 million years ago).

Third, there was fossil evidence, for the same fossil plants and animals are found on multiple continents. As one example, fossils of mesosaurs, a freshwater reptile that lived during the Permian period, are found in South America and Africa. How did these freshwater reptiles come to populate two continents separated by at least twenty-five hundred kilometers (sixteen hundred miles) of saltwater ocean? Glossopteris, an extinct type of fern, provided a plant example. Glossopteris fossils are found in Australia, Antarctica, India, South Africa, and South America. Glossopteris produced large and bulky seeds that could not have drifted or been blown across the oceans to distant continents. The idea of these plants originally evolved on a supercontinent and then spread out as pieces of the continent moved apart best explained the wide dispersion of Glossopteris.

Finally, as a meteorologist and climatologist, Wegener was in a uniquely strong position to assess geologic evidence that seemed to show that most continents in the past had very different climates than they do today. These data might not have been appreciated by a pure geologist lacking training in meteorology and climatology.

The Permo-Carboniferous ice-age era occurred about 280 million years ago. Evidence of this ice age would be expected to be mainly found around the poles. But instead, geological evidence for the Permo-Carboniferous ice age is scattered around the globe, including in areas that today contain the hottest deserts on earth. Saharan Africa bears markings made by ancient glaciers, and conversely, near the North Pole there are remains of tropical vegetation in the form of coal.

Based on Wegener's map of Pangaea, all of the areas showing evidence of the Permo-Carboniferous ice age had long ago been clustered neatly around the South Pole. What are today Africa, Antarctica, Australia, and India had, in Wegener's theory, once composed the southern hemisphere supercontinent now named Gondwanaland. It all made sense if there was continental drift. These climactic observations were the strongest evidence for continental drift.

Objections to the Continental Drift Theory

Geology scholars almost universally rejected Wegener's theory. Their responses ranged from angry and bitter condemnation of the theory itself to ad hominem attacks against Wegener. In one such attack, British geologist Philip Lake accused Wegener of being "quite devoid of critical faculty." In 1911 Wegener wrote a letter to his own father-in-law, Professor Wladimir Peter Köppen, to protest how badly even his own family treated his work: "You consider my primordial continent to be a figment of my imagination."

To the geology community, Wegener was rejecting the very foundations of geological science, theories regarding the formation of the earth that had long been held to be unassailable. Most geologists dismissed his theory as a fairy tale or "mere geopoetry." When Wegener presented it at an international symposium in New York in 1926, many of those in the audience were openly sarcastic to the point of insult. Since there was no way he could convince them, Wegener said little and just sat smoking his pipe and listening. The rejection of his theory was so wide-ranging that even Albert Einstein once wrote a prologue for a book that ridiculed Wegener.

Continental drift was refuted along four lines of reasoning. First, the long-distance movement of giant, massive continents seemed, to say the least, highly improbable. How could such giant land masses simply float around the earth? The very idea defied reason and personal experience.

Second, dispersion of now-extinct species could more plausibly be explained by the existence of ancient, and now long-vanished, land bridges. It seemed far more likely that land bridges had once existed between the continents and that species could have moved along these corridors of elevated land to populate the various continents. The idea that such formations might be fragile and easily lost over time also appealed to common sense. Over many millennia, these land bridges could have been eroded away or submerged by a rise in sea level.

In fact, the current well-accepted explanation for the human habitation of the Americas is that humans arrived there sometime between 27,000 and 14,000 years ago by crossing a now-nonexistent land bridge from Asia. This bridge was formed during the end of the last glacial period, what is called the Last Glacial Maximum, when a giant glacial sheet connected what is now the easternmost portion of Russian Siberian Asia, bounded by the Lena River, to the Mackensie River area of Canada in North America two thousand miles (more than three thousand kilometers) away.

Third, Wegener provided no explanation for the force that moved the continents beyond some nebulous notion that they somehow were dragged over the underlying fluid rock in Earth's mantle. Plus, he had no explanation for how the continents maintained their rigidity as they drifted. At the time, virtually nothing was known about what went on below the surface of the oceans. This was the greatest deficiency of Wegener's theory.

When pressed, Wegener proposed that the rotation of the earth creates a powerful centrifugal force toward the equator. He explained that Pangaea had originated near the south pole and that this centrifugal force caused the protocontinent to break apart and the resulting pieces to drift upward toward the equator. Wegener called this the pole-fleeing force. Similarly, he proposed that the westward drift of the Americas was caused by the gravitational forces of the sun and the moon.

But both of these explanations were quickly and easily rejected. Calculations showed that the forces generated by the rotation of the earth as well as the gravitational forces produced

by the sun and moon were totally inadequate to move continents. Even Wegener came to realize that the forces he had proposed were insufficient to move continents. He wrote in 1929,
"It is probable the complete solution of the problem of the forces
will be a long time coming. The Newton of drift theory has not
yet appeared." The Newton of drift theory did finally appear,
though—thirty-seven years after Wegener's death.

The fourth reason geologists rejected Wegener's theory was
that Wegener was not a geologist, which made his theorizing even
more suspect in the view of the professionals he had to convince.
There is a long history of people making amateurish mistakes
when trying to work in a field outside their own. A recent well-
known example is cold fusion. Nuclear power plants work via
atomic fission. In these plants, an unstable radioactive element
produces a controlled chain reaction resulting in new radioactive
elements and releasing large amounts of heat. Atomic fusion, the
power source of the sun and hydrogen bombs, also releases large
amounts of heat, but in contrast to fission, fusion is based on
nonhazardous hydrogen turning into equally harmless helium.
Sounds like a great source for a power plant.

The problem is that atomic fusion occurs only at temperatures
greater than 100 million degrees Celsius. It is a massive engineering challenge to heat hydrogen to that temperature in order to
initiate atomic fusion. Of course, once started, the resulting heat
produced by atomic fusion could sustain the fusion reaction and
also be harvested to make electricity. If an atomic fusion reactor
could somehow be designed, atomic fusion would be an almost
perfect energy source.

In 1989 two electrochemists from the University of Utah,
Martin Fleischmann and Stanley Pons, announced that they

had designed an apparatus that generated large amounts of heat that they claimed could only have been the result of an atomic process. They also reported that their apparatus produced small amounts of nuclear reaction byproducts, including neutrons and tritium, consistent with their having achieved atomic fusion at room temperature.

Their report was met with a media frenzy. Fleischmann and Pons told journalists that cold fusion could solve environmental problems while providing a limitless, inexhaustible source of clean energy using only seawater as fuel. Money started to pour in to the University of Utah based on expectations that Fleischmann and Pons's discovery would ignite an entirely new and highly profitable industry in the state. But when many laboratories tried to repeat Fleischmann and Pons's experiment, they all failed. Additional studies performed by these outside laboratories revealed flaws and sources of experimental error in the original experiment. These outside laboratories also demonstrated that Fleischmann and Pons had not actually detected nuclear reaction byproducts.

Early twentieth-century geologists were sure Wegener was making the same type of rookie mistakes that doomed Fleischmann and Pons's research. Plus, they did not appreciate a mere meteorologist telling expert geologists how to do their business. Wegener's idea disappeared but was not forever lost.

Plate Tectonics

In the late 1960s, more than a half century after Alfred Wegener first proposed his continental drift theory and more than thirty years after his death, the story of continental drift landed at the Princeton University geology department in the laboratory of a professor named W. Jason Morgan. This was the geology department that occupied the western side of the building where I spent six years earning my PhD.

W. Jason Morgan was a southerner who had completed his undergraduate studies at Georgia Institute of Technology in 1957, majoring in physics. He then served two years in the Navy before enrolling in the graduate school of Princeton University, where he earned his PhD in physics in 1964. After completing his PhD, Morgan decided to switch to fields from physics to geology (actually geophysics) and transferred to the Princeton geology department for postdoctoral studies. He ended up spending virtually his entire career at Princeton.

During his lifetime Morgan often said that his success as a geologist was in no small part the result of having excellent colleagues, men and women whose ideas stimulated Morgan's thinking and with whom he was able to debate his own ideas and theories. As a postdoctoral fellow, Morgan shared an office with an English geologist named Frederick Vine who had come to Princeton from Cambridge University.

Morgan had joined the geology department to study the geology of the moon, which he proposed to do by examining high-magnification photographs of the lunar surface. Vine on the other hand had come to Princeton with the almost exact opposite interest, studying the geology of the ocean floor. Morgan soon became intrigued by Vine's work and switched his own research topic from the moon to deep sea trenches on the sea floor.

After two years of work, Morgan submitted an abstract for the 1967 American Geophysical Union conference titled "Convection in a Viscous Mantle and Trenches." But after submitting his abstract, and only a few months before the start of the meeting, Morgan developed an entirely new research interest. He decided to present his new ideas instead of the work described in the abstract. He began writing notes for his talk just one week before the start of the conference, finishing at 2:00 a.m., just hours before his presentation. Because of the change in topic, when Morgan arrived for his previously scheduled session, he was speaking to the wrong audience. A bunch of geologists who had come to hear about deep sea trenches instead heard a lecture describing an entirely new theory of how continents were formed and evolved, the theory of plate tectonics.

Morgan started off by thinking about the long cracks called fracture zones on the Pacific Ocean floor, features that had been first described by geologists only recently, in the mid-1960s. Morgan calculated that these hairline fractures in the Pacific seafloor all curved around the Earth to a point just north of Siberia. If the fractures had been the result of a local, random geological process, Morgan theorized, they would have occurred randomly. But instead, the fractures were connected with a long-range organization. Morgan thought that the long-range organization of the fractures might be a sign of some significant worldwide process, and he demonstrated mathematically that the fractures had all been formed by a rotation about the same pole. This started him questioning what sort of process could have created the effect.

Morgan's model to explain the long-range structure of the ocean floor fissures, the theory of plate tectonics, said that the earth's surface is composed of fifteen rigid plates. These plates meet each other in a massive, many-thousand-kilometer-long and mostly underwater ridge system, spanning the earth like seams on a soccer ball. He proposed that new ocean floor is created at the mid-ocean plate boundaries. The plates at these edges are warm and thin, allowing hot magma to rise to the surface and spread the ridges outward to form new crust. The new crust then pushes the rest of the plate farther away from the fracture zones.

As the expanding ridges push the plates outward, the rigid plates need to go somewhere. As a result, they crash with great force into the next adjacent plate. The two best-studied mid-ocean ridges are the Mid-Atlantic Ridge and the East Pacific Rise. The Mid-Atlantic Ridge runs down the center of the Atlantic Ocean and is spreading at a rate of 2 to 5 centimeters (0.8 to 2 inches) per year. The East Pacific Rise for the most part runs down the center of the Pacific Ocean and spreads at a rate of 6 to 16 centimeters (3 to 6 inches) per year. As a result of East Pacific Rise spreading, the Pacific Plate collides with the North American Plate and the Nazca Plate collides with the South American Plate on the west coasts of North and South America. This has caused the adjacent plates to buckle and deform, creating the Rocky and Andes Mountain chains. To the west, the collision between the Indo-Australian Plate and the Eurasian Plate has created the Himalayan Mountains.

Plate collisions also generate ruptures in the planetary crust, creating volcanos, openings in the crust where hot lava and gases escape from a magma chamber below the surface. The Ring of Fire, a chain of volcanos along the Pacific coasts of South America, North America and in East Asia on the Kamchatka Peninsula and in Japan, are also the result of the spreading of the plates from the East Pacific Rise under the Pacific Ocean. In the same regions, the energy from the plate collisions is periodically abruptly released, producing strong earthquakes.

Morgan's plate tectonics theory confirmed every aspect of Wegener's theory of continental drift. In addition, plate tectonics also addressed the greatest weakness in Wegener's theory, the source of the force that moved the continents. Wegener's continental drift arises from the huge forces generated by Morgan's spreading of the tectonic plates.

Notwithstanding the graduate student's bumper sticker, Gondwanaland will never be reunited. But plate tectonics did restore Wegener's scientific reputation. Instead of being portrayed as a novice dabbler in the science of geology, Wegener is now regarded as a brilliant visionary who was able to understand things that his contemporary geologists were blind to.

Today Wegener is considered to be one the greatest geologists of all time. The Alfred Wegener Institute for Polar and Marine Research in Bremerhaven, Germany, was established in 1980 on the hundredth anniversary of Wegener's birth and awards the Wegener Medal in his honor. Also honoring him are the Wegener craters on both the moon and Mars, the asteroid 29227 Wegener, and the Wegener peninsula in Greenland, the place where he died during his last polar expedition.

The stories of Alfred Wegener's theory of continental drift and biologist Peyton Rous's discovery of Rous sarcoma virus have many similarities. Early in their careers both scientists promoted what are now recognized to be correct theories, yet when originally proposed their ideas were unprovable due to technical limitations of the time. Rous was an experimentalist who lacked the technical tools to definitively solve his problem. Wegener, who was not an experimentalist, lacked the necessary detailed knowledge of his system that would only be established thirty-five years after his death at a relatively young age. Rous was more fortunate. Living to the age of ninety, he witnessed the tools needed to prove his hypothesis being developed during his lifetime, and thus he became the oldest physiology or medicine Nobel Prize recipient, receiving his award at the age of eighty-six, just four years prior to his death.

14 Robin Warren and Barry Marshall
Discovered the cause of and treatment for stomach ulcers

Ulcers

Stomach ulcer is a common disease that was well known in antiquity. Approximately 10 percent of us will at some point during our lives suffer from a stomach ulcer. The Roman emperor Marcus Aurelius was a famous sufferer who died in 180 AD from a perforated stomach ulcer. His doctor, the prominent Roman physician, Galen, was powerless to save his patient.

The interior of the stomach is a hostile and toxic environment. The cells of the stomach lining produce copious amounts of hydrochloric acid, a highly corrosive mineral acid, to aid in the digestion of our food. The chemical measure of acidity is pH. The lower the pH number, the more acidic a substance is. The stomach is extremely acidic, with a pH between 1 and 2. Battery acid is pH 0, lemon juice is approximately pH 2–3, acid rain is pH 4.3, blood is roughly neutral at pH 7.4, and sea water is pH 8.1. Thus, the interior of the stomach is not too terribly different from battery acid.

It was common sense to expect that this extreme level of acidity could sometimes injure the stomach. In 1915 a physician scientist named Bertram Sippy devised a treatment that came to be known as the *Sippy regimen*. His idea was a simple application of high school chemistry: neutralize the acid with base. Patients would be given basic foods, such as milk and cream, once an hour with the gradual addition of eggs and cooked cereal in combination with alkaline powders. The Sippy regimen provided symptomatic relief for many patients. But soon it became clear that this treatment also produced a serious and sometimes fatal toxic condition in patients that came to be known as milk alkali syndrome.

Milk alkali syndrome is the result of the high calcium and high pH levels in the blood that are produced by the Sippy regimen diet. Symptoms include poor appetite, dizziness, and headache and, in extreme cases, confusion, psychosis, kidney failure, and death. Thus, not highly effective and yet often dangerous, the Sippy regimen was soon abandoned.

Treating ulcers with high school chemistry was not going to work. A different approach was needed. In the 1960s a group of scientists at the pharmaceutical company Smith Kline & French (SKF) came up with a new idea. Instead of neutralizing the acid, they proposed to find a drug that would block the acid from being formed in the first place.

One of the SKF scientists, Robin Ganellin, worked in close collaboration with consultant James Black on the effects of the neurotransmitter histamine. Histamine was known to stimulate the secretion of acid in the stomach, but antihistamines, drugs like Benadryl that had been around for twenty years, did not block stomach acid secretion. Ganellin and his collaborators reasoned that perhaps the receptor that mediated the action of histamine in the stomach was different from the receptor that the conventional allergy-treating antihistamines worked on. If true, they reasoned, it should be possible to design a special type of antihistamine to block the stomach receptor and thus block acid production in the stomach.

The SKF scientists called the drug they eventually discovered cimetidine. It took twelve years of research effort, was approved in the United Kingdom in 1976 and the United States in 1979, and was sold under the brand name Tagamet. Tagamet worked, and it worked well. James Black received the 1988 Nobel Prize in Physiology or Medicine in part for his work on cimetidine. Tagamet became the first pharmaceutical blockbuster drug, the first drug in history to produce more than $1 billion annually in sales. This discovery soon led other pharmaceutical manufacturers to come up with their own versions of the drug: ranitidine (Zantac) by Glaxo Pharmaceuticals and famotidine (Pepcid) by Yamanouchi Pharmaceuticals.

These so-called histamine H2 blockers (named after the H2 stomach histamine receptor they acted on) were highly effective but imperfect drugs. As a result, they inspired scientists to try to move to a next stage of therapy by finding something even better. If shutting off the system that regulated acid production at the histamine level worked well, it stood to reason that a drug working downstream from the H2 receptor could be even better. Tagamet blocked cells in the stomach from responding to histamine to produce acid. But what actually made the acid in the first place was a protein called the proton pump, a molecular pump that moved acid from the inside of the cells lining the stomach and literally pumped it out into the interior of the stomach.

The stomach proton pump was discovered in the late 1970s, and the pharmaceutical company AstraZeneca produced the first proton pump inhibitor, omeprazole (Prilosec), a drug that was clearly superior to cimetidine. A single dose of cimetidine produces a considerable lowering of stomach acid for about twelve hours. But omeprazole will produce at least a similar, and often a far greater effect, and lasts for more than sixteen hours. Prilosec is far more effective than Tagamet in treating stomach ulcer, plus it treats another disease that is caused by stomach acid: gastro-esophageal reflux disease, commonly called GERD.

In GERD, stomach acid rises up into the esophagus. The acid produces an unpleasant taste in the back of the mouth, heartburn, bad breath, and in some patients, chest pain, breathing problems, and damage to tooth enamel. GERD damages the esophagus over time and in the most extreme cases can cause Barrett's esophagus, a premalignant condition that will transition to esophageal adenocarcinoma, an often deadly cancer.

About 20 percent of adults have GERD. In order to provide relief from GERD, a drug has to lower stomach acid to a far greater degree than is needed to treat a stomach ulcer. Proton pump inhibitors are effective in treating GERD, but H2 blockers are not.

AstraZeneca filed an investigational new drug application for omeprazole in 1980, and the drug was taken into phase III human trials in 1982. Omeprazole was launched in the US in 1989 under the trade name Prilosec and by 1996 had become the world's biggest selling pharmaceutical. More than 800 million patients worldwide had been treated with the drug by 2004. The commercial success of omeprazole encouraged competing pharmaceutical companies to develop similar, rival medicines including lansoprazole (Prevacid, launched in 1995) and pantoprazole (Protonix, launched in 2000).

It thus appeared that the problem had been solved. Physicians were sure that high stomach acid levels caused stomach ulcers, and they now had safe and effective medicines to treat the condition. Most scientists moved on to study other diseases, but not everyone.

Robin Warren

Robin Warren's ancestors were early English settlers who had moved to Australia during the first half of the nineteenth century as part of a land investment program. The family had a long tradition of serving as medical doctors, and many of Warren's ancestors were prominent physicians in the Australian city of Adelaide. One of his uncles was a member of the Australian Army Medical Corps during World War II.

Warren was a good student and considered following the family tradition by studying medicine. But he had developed epilepsy as a teenager, and his condition was only partially controlled by medication. Though many people thought that having epilepsy would preclude a medical career, with his parents' support Warren entered Adelaide University Medical School in 1955. He took his clinical training at Royal Adelaide Hospital, which at the time was the only general teaching hospital in Australia.

After medical school, Warren became a junior resident medical officer, the equivalent of a resident in the United States. He had particularly enjoyed his medical school anatomy classes and decided to specialize in pathology. His first position was as a temporary lecturer in pathology at the Adelaide University, where he learned morbid anatomy and histopathology. From there he moved to the Royal Melbourne Hospital, where, working under the tutelage of David Cowling and Bertha Ungar, he passed his exams in hematology and microbiology to become a fully fledged pathologist.

Warren, now fully credentialed, took a position as a pathologist at the University of Western Australia and the Royal Perth Hospital in Perth, Western Australia. The custom in eastern Australia was for pathologists to become generalists, to master a broad spectrum of pathology knowledge and practice. In contrast, the custom in Western Australia was for pathologists to specialize early on. As a fortunate result of his early training, Warren had much broader knowledge and interests than his Perth colleagues, a perspective that would make an important contribution to his later work.

The gastrointestinal endoscope allows physicians to examine the interior of the stomach and take samples when needed. The first practically usable endoscope was designed in 1959, and during the next twenty years gastrointestinal endoscope instruments were developed for use in routine clinical medicine. As a pathologist, Warren was often called upon to evaluate the biopsies that had been collected via this new technique. In 1979 Warren noticed bacteria growing on the surface of gastric biopsies.

When a new analytical technique is introduced into medical practice, no one knows what they will find. If something unusual and unexpected is seen, one must consider multiple possible explanations. In this case, the first possibility was that the bacteria Warren had seen were some sort of contaminant introduced by him when he was preparing the samples. Everyone knew that the high level of stomach acid would make it impossible for any type of microorganism to grow there. The second possibility was that what he observed may have simply been normal, but just never previously seen. Without the endoscope, samples of the type he had been examining could not have been obtained from patients. And the third possibility was that the unexpected observation of bacteria in a gastric biopsy was associated with a disease. And since Warren saw the bacteria in many samples, if they were associated with a disease, it would have to be a very common one.

Warren soon became intrigued with these bacteria. First of all, they seemed to be a new, previously unstudied species. If they were a contaminant, they would have had to come from someplace in the laboratory, but common laboratory bacteria had been studied for decades and were all well known. If the bacteria were not a contaminant, what was their significance? All his biopsies came from sick people whom the hospital was trying to treat. In order to figure out if the bacteria had any medical significance, Warren had to get biopsies from healthy people to see if they also had the bacteria. But having a gastrointestinal endoscopy is no fun. Convincing healthy people to submit to one just so he could check out his pet idea seemed an unlikely prospect. Normal biopsies from the gastric antrum were very rare, but by persisting, Warren eventually collected twenty normal biopsy samples. None showed the bacteria.

Warren began to prepare a paper for publication to report his findings. He recruited a young physician named Barry Marshall to help him gather clinical history data, arrange for improved biopsies, collect specimens for culture, and catalog the endoscopy findings that were needed for his publication. Together they intended to report finding a new bacterial species in the human stomach and to link infection with this new bacterium to duodenal ulcer disease.

At the time, this seemed nuts. Everyone knew that a stomach ulcer was not an infectious disease. No one calls their boss on Monday morning to say, "Sorry, I can't come in today. I caught an ulcer over the weekend." Warren's wife, who was a physician specializing in psychiatry, helped him by proofreading his papers and making editorial suggestions for revision. In Warren's Nobel Prize autobiography, he wrote of her: "Particularly as she was a doctor, and knew the standard teaching that nothing grows in the stomach, and therefore that I was trying to prove the 'impossible.' As a psychiatrist, she could have suggested I was mad."

If Warren's supportive and protective spouse could have ventured that suggestion, it's easy to imagine the reaction of the scientific community to the two papers that Warren published, one in 1983 and a second paper with Marshall in 1984, both in the medical journal *Lancet*. Everyone thought they were crazy, and in hindsight it's remarkable that the editors even accepted the paper for publication in the first place. *Lancet* is a highly respected and conservative medical journal, considered to have extremely high publication standards. But fortunately, despite their highly unconventional hypothesis, the extremely meticulous way in which Warren and Marshall planned and conducted their research led to their papers' acceptance.

Barry Marshall

Barry Marshall's family lived in Western Australia. His grandparents ran a hotel in Kalgoorlie, a poor working-class town where young men would typically leave school at the age of sixteen to work in the mines. But Barry Marshall's parents had higher aspirations for their children, and early in his life his family moved to Perth, where it was hoped there would be better professional opportunities.

Marshall was an excellent student, interested in all sorts of science and technology. After high school he chose to study medicine because he thought it would require fewer of the mathematics courses he disliked than other technical fields. After graduating with a medical degree from the University of Western Australia in 1975, Marshall performed his internship and residency studies at Queen Elizabeth II Medical Centre. In 1979 he moved to Royal Perth Hospital for specialty training. As part of this training, Marshall was encouraged to perform a clinical research project each year. In 1981 during his rotation in the gastroenterology division, he met Robin Warren, who needed help with his research.

Warren gave Marshall a list of patients with funny curved bacteria present in their stomach biopsies and assigned him to follow up with them to learn what clinical diseases they had. One of the patients on Warren's list was a woman Marshall had earlier seen in the wards. She had severe stomach pain but no diagnosis. Her physicians could find no physical explanation for her gastric pain and referred her to a psychiatrist, thinking her complaint was psychosomatic. Today, based on Warren and Marshall's research findings, the redness in the patient's stomach and the presence of Warren's bacteria in the stomach biopsy would lead to a clear, unmistakable, definitive diagnosis of stomach ulcer. But at the time no one thought this way.

Marshall wasn't a gastroenterologist and therefore wasn't bound by their ideas, such as the presumption that bacteria couldn't survive in the highly acidic stomach. Having been trained as a generalist, he also was open to Warren's unconventional thinking. He quickly became hooked by Warren's hypothesis, started reading the published literature on the subject, and continued to collaborate with Warren on the research project long after his gastroenterology rotation was over.

Warren and Marshall's Research

Following up on Warren's observation of bacteria in patients' stomachs, the first question was: What type of bacteria were these? The bacteria initially appeared to be a new species of bacterium related to an obscure group of germs called Campylobacter. Campylobacter can spoil food, especially poultry, and was known to cause mild food-borne illnesses.

But the bacteria that Warren found weren't an exact fit to be included in the Campylobacter group. Warren and Marshall decided their stomach bacteria likely belonged to a new group, which they called *Helicobacter*, named after the bacteria's helical shape under the microscope. Needing a species name for their organism, they decided to call it *pylori* after the bacteria's preferred habitat; the pylorus, the point where the human stomach attaches to the intestine. Thus the scientific name for the new species became *Helicobacter pylori* or, for short, H. *pylori*.

The second question was: What were the bacteria doing there, and why were they predominantly, if not exclusively, in the stomachs of ulcer patients? There were two possibilities. Perhaps ulcers made the stomach more hospitable to the H. *pylori* bacteria. The bacteria were possibly just present in the pylorus, free-riding out the peptic ulcer disease. A second, and more exciting, possibility was that the H. *pylori* was actually causing ulcers. Fortunately, Warren and Marshall had a widely accepted standard means to show whether or not a germ causes a disease.

Prior to the mid-nineteenth century, the predominant idea for what caused what we now call infectious diseases was the miasma theory. The miasma theory stated that diseases like cholera, smallpox, and the Black Death were caused by miasma, a noxious form of "bad air," also called "night air." The air at night was considered dangerous in most Western cultures, and people feared going out after dark would expose them to disease.

But by the mid-nineteenth century, the germ theory was proposed. Microorganisms like bacteria and fungi were first discovered in the mid-1600s. Scientists during the next two hundred years studied these organisms and showed how they caused things that we see in everyday life, like bread rising, alcohol fermentation, and food spoiling. The new germ theory stated that microorganisms, tiny living things that can be seen only with a microscope, can cause disease by infecting people and then growing and reproducing within them. At the time, this idea was hard for many people to accept. It seemed highly counterintuitive. How could a tiny, invisible thing be powerful enough to sicken or even kill an adult human being?

But with time this idea eventually took hold. It soon became clear that a broadly accepted set of criteria was needed to clearly define what was, and what was not, an infectious disease. The late-nineteenth-century German physician scientist Robert Koch was one of the founders of modern microbiology. Koch identified the pathogenic bacteria that caused many important infectious diseases, including tuberculosis, cholera, and anthrax, and he was awarded the Nobel Prize in Physiology or Medicine for his work in 1905. He was a major supporter of the germ theory of disease and created a system to identify infectious diseases, a system that today is called Koch's postulates. Koch's postulates, which prove whether or not a disease is an infectious disease, state:

1. The proposed causative microorganism must be found in all individuals suffering from the disease but should not be found in healthy individuals.
2. The microorganism must be isolated from a diseased individual and grown in pure culture.
3. The cultured microorganism should then cause disease when introduced into a healthy individual.
4. The microorganism must then be reisolated from the inoculated, diseased experimental host individual and shown to be identical to the original specific causative agent.

Warren and Marshall had fulfilled postulates 1 and 2 for *H. pylori* and stomach ulcers. They had cultured *H. pylori* from ulcer patients and not from healthy patients. They had grown and characterized *H. pylori* in pure culture. But it was clear that if they wanted to convince the world that *H. pylori* caused stomach ulcers, they needed to also fulfill postulates 3 and 4.

Marshall tried to develop an animal model for *H. pylori* infection so he could determine if infection resulted in stomach ulcer development, but he was unsuccessful. In 1984 he instead followed an alternate and highly controversial route. Marshall decided to use himself as a guinea pig, the animal model. He did not discuss his decision with the ethics committee at the hospital as he properly should have done, nor did he tell his wife, because he was certain she would forbid him from intentionally infecting himself with a germ he believed to be a pathogen.

Marshall drank a "brew" consisting of a suspension of two culture plates worth of *H. pylori*. After five days he started to have bloating and fullness after the evening meal, and his appetite decreased. His breath became bad and he vomited clear watery liquid early each morning. A follow-up endoscopy showed severe active gastritis with damage to the stomach lining. The *H. pylori* had not simply passed through his digestive tract. Instead, Marshall was starting to develop an ulcer. His wife insisted that Marshall immediately start treating himself with antibiotics, and he quickly recovered. Koch's postulates had now been fulfilled, proving that stomach ulcer was an infectious disease. Marshall later published a paper describing these experiments with a "male volunteer," but it soon became widely known that the male volunteer was none other than Marshall himself.

Robin Warren was one of the first people Marshall told about his self-experiment. Shortly thereafter, Warren received an early morning telephone call from an American journalist who had miscalculated the time difference and called at 5:00 a.m. Perth time. The journalist wanted to know why Warren was so convinced that *H. pylori* caused ulcers while most of the medical community believed otherwise—that *H. pylori* was a harmless bacterium that sometimes grew in the human stomach. Warren was caught off guard, and who would not be at 5:00 a.m.? He blurted out, "I know because Barry Marshall has just infected himself and damn near died." It was an exaggeration, but it was a great quote for the journalist to include in his story. Conventional broadsheet newspapers didn't consider Warren and Marshall's work to be of sufficient interest or credibility to merit publication, but this reporter worked for the notorious tabloid *Star*. The next day Warren's story appeared: "Guinea-pig Doctor Discovers New Cure for Ulcers . . . And the Cause."

Self-Experimentation

Self-experimentation is far more common in medical science than is generally appreciated. Barry Marshall wrote that he was inspired by the story of John Hunter, an eighteenth-century British surgeon who infected himself with gonorrhea and syphilis as part of his research studies, which likely caused his death just a few years later.

During the 1980s I was working on an antibiotic project. We were trying to develop a version of an injectable antibiotic that could instead be taken in pill form. We had a lead compound that was orally active in animals, but we knew that oral activity experiments in animals were not always a reliable predictor of oral activity in humans. Being unsure of the compound's human oral activity, we were hesitant to start expensive formal human trials with the compound. Therefore, several of us volunteered, likely illegally, to take the experimental drug.

It was a simple experiment. We took the drug by mouth as soon as we arrived at work. We knew that the drug was eliminated from the body in the urine, so if the drug was orally absorbed it should appear in the urine in the late morning. We all gave urine specimens at around 11:00 a.m., and a quick test showed that the drug was in our urine. We were elated. The drug was orally absorbed in humans.

This had been a particularly busy day for me, and I had skipped lunch. But by midafternoon I was really hungry and noticed the remains of a buffet in the hallway left over from a lunchtime seminar. I went over and grabbed a scoop of chicken salad, some bread and a cola and went back to my desk to work while I ate lunch.

Late that evening after dinner, I suffered a bout of wrenching diarrhea. My immediate thought was that I was suffering from food poisoning. The chicken salad with which I had made my sandwich had stood out at room temperature for at least three hours, probably far longer. Chicken and mayonnaise is a superlative medium for the growth of Salmonella bacteria, the most common cause of ptomaine poisoning. And the timing was perfect; about a ten-hour incubation period before symptom appearance. The next morning I was fine and never gave it another thought, that is until months later when we started clinical trials. Multiple volunteers developed severe diarrhea after taking our test compound. It was the experimental drug that had caused my diarrhea. Needless to say, that drug never made it to clinical approval.

The Long Road to Acceptance

Warren and Marshall started off as total outsiders. They were practicing physicians, not scientists, and they were not even gastroenterologists. They had proposed a hypothesis that not only was the opposite of conventional thinking in the field but that conventional medical wisdom held to be impossible. Even Warren's wife, a fellow physician, had major reservations about accepting what he had proposed.

But then, with time, many scientists around the world were successful in repeating Warren and Marshall's findings. These scientists soon came to believe that even if Warren and Marshall's newly found bacteria didn't cause disease, these bacteria were not very difficult to find in the stomach. One early source of support was scientists in the Campylobacter scientific community. This relatively obscure organism, Campylobacter, was studied by a small, close-knit group of investigators. They were enthusiastic about the new findings on an organism so closely related to their own and were eager to replicate and extend Warren and Marshall's observations.

Another source of support was Mike Manhart, a microbiologist working for Procter & Gamble. Procter & Gamble sold Pepto Bismol, an over-the-counter stomach remedy that contained bismuth as its active ingredient. It had been known for years that ulcer patients could obtain relief from their illness by taking bismuth, but no one knew why. The activity of Pepto Bismol in ulcer patients now made total sense. Bismuth is extremely toxic to *H. pylori*, so the effectiveness of bismuth in treating ulcer patients could be easily explained if ulcers were caused by *H. pylori* and bismuth treated the disease by killing the disease-causing pathogenic *H. pylori* bacteria. Manhart saw much economic potential for Procter & Gamble if it could be shown that Warren and Marshall were correct and thus provided them with significant support, both scientific and financial.

Initially, Warren and Marshall's papers were rejected for publication, and even the accepted papers were delayed significantly by editorial disputes prior to publication. There was constant criticism that their conclusions were premature and not well supported. When the work was presented at conferences, the results were hotly debated and disbelieved, often on the basis that the results simply could not be true. The bacteria Warren and Marshall had described were said to be either contaminants or highly prevalent harmless bacteria that were commonly associated with humans.

The tide of acceptance began to turn in the early 1990s. At meetings Marshall would attend, he began to receive as much praise as criticism. In February 1994, the American National Institutes of Health held a consensus meeting in Washington, DC, and after two days released a statement saying that the key to treatment of duodenal and gastric ulcer was the detection and eradication of *Helicobacter pylori*. In 2005 Warren and Marshall won the Nobel Prize in Physiology or Medicine. Today the standard treatment for stomach ulcers is a course of treatment with antibiotics, drugs that cure the disease rather than merely suppress symptoms. These antibiotics were discovered in the 1950s, but everyone thought they would be worthless for the treatment of stomach ulcers, and as a result, no one tried.

It may seem unusual that Barry Marshall engaged in self-experimentation in order to prove what he had discovered. But, as explained earlier, self-experimentation in medical science is not as uncommon as is generally believed. Ethicists write that that there are no ethical barriers to self-experimentation. When the self-experimenter is the actual investigator, there is no one in need of protection from unethical behavior. At least fourteen Nobel laureates have engaged in self-experimentation. Seven of them used self-experimentation during the course of the investigations for which they won the Nobel Prize. Barry Marshall was one.

Epilogue

Epilogue

Recognition for scientific accomplishments is commonly delayed, and the vast majority of important scientists become famous only late in their lives or, not uncommonly, after their deaths. But there are two types of scientific discoveries. The types discussed in this book are discoveries that change the way we think about how the world works. For such discoveries, there is always an entrenched faction of scientists who are dedicated to an old way of thinking and oppose the new idea. Max Planck argued that members of this group need to die off before the new idea is accepted.

The other type of discovery is research methods, tools that accelerate new scientific discovery. Recognition of research method discoveries works very differently. After a new research method is published, many scientists will try it out to see if it is useful in their own investigations. And if it is useful, there will soon be a large number of reports from a diverse group of laboratories describing and supporting the value and benefits of the new method. A few examples:

Monoclonal Antibodies

A method for the study of the immune system and for
the production of unique assays and new medicines

When the immune system responds to an infection, it creates new antibodies to fight it. But the immune system doesn't make just one type of antibody; it makes thousands of different ones to fight the infection. Some of them work well, and some work poorly. The point is that the serum of a person responding to an infection contains a witch's brew of different antibodies, all of which have at least some effect against it.

The fact that all antibody responses are mixtures left open a number of questions about antibodies and how they work. What is the detailed structure of an antibody? What causes the difference between high-efficacy antibodies and low-efficacy antibodies, and how does this relate to structure? Why do some antibodies act against one pathogen, while others act against a different pathogen? It was a hopeless task to attempt to purify individual antibody types from these mixtures to address these questions.

And there was no way to produce pure antibodies, with the exception of cells from a rare cancer called multiple myeloma. In multiple myeloma, one specific antibody-producing cell becomes cancerous, grows uncontrollably, and pumps out huge amounts of a single, pure antibody. But multiple myeloma cells cannot be created on demand. A chance event produces these cells, and as a consequence one would have the antibodies but not know the antigenic target they acted on. Multiple myeloma researchers had the lock, but not the key.

In 1975 a British-Argentine scientist named César Milstein published a method by which individual antibody-making immune cells could be made immortal. Once immortal, these cells could then serve as a factory to make large, pure amounts of a desired antibody. Such antibodies are called monoclonal antibodies.

In addition to answering many important basic questions in immunology, this method enables an enormous number of applications. Among them has been the development of a wide array of cheap, sensitive, and highly reliable medical tests and the production of antibodies for use as medicines. Today about one-third of all new FDA-approved medicines are monoclonal antibodies. Milstein was awarded the Nobel Prize in 1984, only nine years after he first published his monoclonal antibody method.

PCR (Polymerase Chain Reaction)
A method to synthesize a large quantity of DNA sequence starting with a single DNA molecule

The story of the discovery of the polymerase chain reaction method, commonly called PCR, is similar to that of the monoclonal antibody. By the late 1970s and early 1980s, the application of genetic engineering methodology had turned into a major business. But a significant technical limitation of that era was the inability to make sequences of DNA by chemical synthesis. The DNA sequences that were needed by researchers in the field at the time were of a size and complexity that rendered them well beyond the capacity of conventional organic chemical synthesis.

In the early 1980s, Kary Mullis was a chemist working for Cetus Corporation and was assigned to synthesize DNA sequences for use in their various projects. Because conventional organic chemical synthesis methods are not capable of synthesizing the desired large, complex molecules, he instead conceived of the polymerase chain reaction, now called PCR, a totally new process that was able to do the job.

In PCR you start with a tiny amount of the DNA containing the sequence of interest. This sequence can be either pure or part of a mixture. That is, you can start with an entire genome and PCR just one single gene from a mixture of thousands of genes. You then mix the DNA sequence with a cellular enzyme called DNA polymerase. This is the enzyme that makes new DNA in dividing cells so that each of the resulting cells following division has a full complement of DNA sequences. Then you add two extremely short stretches of synthetic DNA sequence, ones that are small enough to be made by conventional chemical synthesis methods. These are called primers, and they encode sequences at the opposite ends of the target DNA sequence of interest. When this mixture is incubated, the DNA polymerase races back and forth between the two primers copying the DNA sequence in between.

The reaction works geometrically. Say you first have one molecule to copy. After the first round, you have two, then four, then eight, etc., until after a short period of time you have a substantial amount of the target DNA sequence.

In addition to being an incredibly powerful research tool, PCR has multiple commercial applications. The most reliable tests for COVID-19 disease are based on PCR. Forensic law enforcement methods depend on PCR to amplify DNA sequences left at the crime scene. PCR is used in diagnostic procedures to detect disease mutations in patients. It has enabled the study of human evolution by amplifying tiny amounts of DNA in ancient human fossil skeletons. It also provides the basis for contemporary paternity testing.

Mullis conceived of PCR in 1983. With help from other Cetus scientists, PCR was reduced to practice—in other words, shown to be workable in solving real-world problems—in 1986, and in 1993, only seven years later, Mullis was awarded the Nobel Prize.

Mullis was not a typical scientist. Unlike the ordinary scientist who labors his or her whole life trying to solve a research problem, Mullis was incredibly mercurial. He clandestinely synthesized LSD and other hallucinogens in the lab while working on his PhD. He was hired by Cetus in 1979 but quit in 1986, just when his idea for PCR was proven out. After quitting, he spent his time surfing and playing the guitar while working intermittently. At one point he started a business to sell pieces of jewelry containing the PCR-amplified DNA of deceased famous people including Elvis Presley and Marilyn Monroe.

Genome Editing
A method to make custom DNA changes in a living organism

A third and more recent example is the work of French microbiologist Emmanuelle Charpentier and American biochemist Jennifer Doudna, who developed a new method for gene editing.

Advanced animals have two immune systems to defend against pathogens: innate immunity and adaptive immunity. The innate immune system is extremely old and present in almost all multicellular organisms: vascular plants, insects, and animals. It works by producing broadly toxic substances in reaction to an infection, substances that are toxic to the animal but more toxic to the invading pathogen. Much of the reason you feel sick when you have an infection is due to the toxic effects of innate immunity.

Later on in evolution, a second immune system appeared: adaptive immunity. This second system produces cells and antibodies specifically designed to fight the infection. The value of this system comes from the fact that it is extremely selective. The antibodies and white cells that are produced by adaptive immunity in response to invasion by a pathogen will kill that pathogen and that pathogen alone. These antibodies and cells are harmless to host cells and other pathogens. Adaptive immunity is an enormous improvement over the older innate immune system.

The thinking by the scientific community had been that only highly evolved multicellular organisms carry an adaptive immune system. But in the late 1980s three laboratories independently discovered that many bacterial species have a special type of sequence in their DNA, which was named CRISPR, an acronym for *clustered regularly interspaced short palindromic repeats*. The scientists asked why these sequences were present in the genomes of bacteria and what they did.

Investigations by multiple laboratories revealed that the CRISPR sequences were part of an adaptive immune system in bacteria and one that works by an entirely differently mechanism from our own. The CRISPR system recognizes foreign DNA when it appears inside the cell (for example, DNA from a virus infecting a bacterium) and mobilizes an enzyme system to specifically chop up the foreign DNA into tiny pieces, protecting the bacterium from the infection.

This was an important new finding about how life works. As the details of how the CRISPR system operates were revealed, some scientists began to wonder whether this process could be adapted to help solve scientific research problems and to treat diseases.

The ability to make precision modifications to the genomes of living animals had long been an elusive goal of the molecular biology community. But now there was a way to do it. Charpentier and Doudna realized that they could exploit components of the CRISPR system to produce a method by which one specific base pair, a single pair within the human genome of 3 billion base pairs, could be changed selectively in exactly the way that the experimenter desired. Their invention is called CRISPR gene editing.

The two scientists published their method for genome editing in 2012, and they were awarded the Nobel Prize in 2020, just eight years later. Applications of CRISPR currently being pursued by the research community include the development of animal models of human diseases, methods to cure inherited diseases (such as sickle cell disease, hemophilia, and cystic fibrosis) by eliminating the mutation that causes the disease, the development of new treatments for infectious diseases, and cell-based immunotherapies for the treatment of cancer.

Danish physicist Neils Bohr proposed that the goal of science is not to reveal universal truths. Rather the goal of science is "the gradual removal of prejudices." The discovery that the sun sits at the center of our solar system counters the prejudice that the earth is the center of the universe. The discovery of microbes counters the prejudice that disease is a punishment from God or caused by something vaguely known as "bad air." And the theory of evolution opposes the prejudice or hubris that humanity is a special and unique creation.

Innovation is tough. One should be compassionate toward creative people in all fields of endeavor. Their contributions to the human condition are of extraordinary value, while at the same time creative people often suffer deeply in their struggle to achieve validation and recognition.

Photo Credits

1. **Max Planck**: Wikimedia Commons.
2. **Gregor Mendel**: Wikimedia Commons.
3. **Barbara McClintock**: Courtesy of Smithsonian Institution Archives, Accession 90-105, Science Service Records, Image No. SIA2008-5609.
4. **Galileo Galilei**: Justus Sustermans, Portrait of Galileo Galilei, National Maritime Museum, Florence; Wikimedia Commons.
5. **Ignaz Semmelweis**: Portrait of Ignaz Semmelweis, aged 42; Wikimedia Commons.
6. **Peyton Rous**: Getty Images, Huron Archive, Keystone/stringer.
7. **Roger Revelle**: Photo by Wade Baker, UCSD Special Collection.
8. **Rachel Carson**; U.S. Fish and Wildlife Service, Wikimedia Commons.
9. **Stanley Prusiner**: Photo by Mark Wilson/Getty Images.
10. **Amedeo Avogadro**: Wikimedia Commons.
11. **David Cushman and Miguel Ondetti**: The Lasker Foundation.
12. **Surendra Nath "Suren" Sehgal**: Courtesy of Ajai Sehgal.
13. **Alfred Wegener**: Wikimedia Commons.
14. **Robin Warren and Barry Marshall**: Robin Warren: Akshay Sharma/Wikimedia Commons; Barry Marshall: China Photos/Getty Images.

Appendix
Struggling for Recognition in the Arts:
Robert Johnson
*Virtuoso guitarist and musical pioneer
of the blues musical style*

In this book, I have discussed how Planck's principle has applied to great scientists, but I don't mean to imply that only scientists suffer the frustration of failing to get truly visionary or breakthrough work accepted. Creative people in many if not all fields endure similar problems. A few well-known examples:

Edgar Allan Poe, who was born in Boston, Massachusetts, in 1809, is known to have written seventy poems, sixty-six short stories, nine essays, one complete and one incomplete novel, and an unfinished play during his lifetime. The total earnings from his literary efforts was $6,200 spread over twenty-plus years, equaling about $10,000 per year in 2023 dollars. Beset by a long list of personal problems, Poe eventually became discouraged and disheartened. He was found wandering around Baltimore, delirious and disheveled, on October 3, 1849, and he died four days later at the age of forty. Although Poe achieved some limited acclaim while he was alive, today he is revered and considered to be one of the most important writers of nineteenth-century America and the inventor of the modern detective story.

Franz Kafka, born in 1883, lived in Prague during the late nineteenth and early twentieth centuries and wrote three novels and seventy-five short stories, some of which he included in three story collections. Most of Kafka's literary output wasn't published during his lifetime. What little he did publish appeared in obscure literary magazines and received little critical approval or positive recognition. Discouraged, when he was dying of tuberculosis at the age of forty-one, he asked his best friend, Max Brod, to burn the extensive collection of his unpublished writings. But Brod failed to carry out Kafka's wishes. Instead, he promoted his friend's published work and brought his unpublished writings into print. Kafka finally began to achieve the fame he deserved in the 1960s, forty years after his death. Today Franz Kafka is a cultural icon, virtually all major universities offer courses on his writing, and literary scholars rate him as one of the greatest writers of the twentieth century.

Vincent van Gogh lived during the second half of the nineteenth century. Judging from accounts, very few people liked his unusual style of painting. When he committed suicide in 1890 at the age of thirty-seven, van Gogh had been able to sell only one painting of the more than nine hundred paintings and eleven hundred drawings and sketches he produced during his lifetime. The painting he sold brought the sum of approximately $109 in today's dollars. Now van Gogh is one of the most famous painters ever to have lived and is considered to be one of the most influential figures in the history of Western art. His painting *L'Allée des Alyscamps* was sold in 2015 at an auction at Sotheby's New York for $66.3 million. On a more macabre note, in 2019 the revolver van Gogh used to commit suicide was purchased by an enthusiast for $182,000.

I am an amateur jazz musician. The great early twentieth-century blues master Robert Johnson therefore holds special interest for me. Johnson died in obscurity in 1938 at the age of twenty-seven. He lived during a time when the blues was first developing as a musical form. In the absence of an established catalogue of blues styles and techniques, Johnson was forced to become immensely innovative, creating and refining his art for much of his short life as an itinerant musician in the Jim Crow South. His work was discovered more than a quarter century after his death and has been highly influential ever since. By way of a conclusion to this book and as a final example of a nonscientist achieving recognition only belatedly, in this case well after death, I provide a brief account of, and tribute to, Robert Johnson.

The Blues

The blues is a uniquely American style of music that emerged from the American Deep South during the late nineteenth century. Based on African musical traditions, African American work songs, and African American spirituals, the earliest blues song lyrics followed a call and response format around stories of personal woes, the harshness of life, lost love, and generally hard times, from poverty to inequality, racial discrimination, bigotry, and many such ills.

Musically, the blues soon settled into a repetitive twelve-bar AAB pattern, now the well-known blues chord progression. Typically, the first "A" section of four bars is based on the first, tonic chord of the scale. The second four-bar "A" section largely repeats the first A section, but instead starts with the fourth chord of the scale and then modulates back to the tonic one chord. And the final four-bar "B" section starts with the fifth chord, then modulates to the fourth chord and finally resolves to the first tonic chord. For example, many blues songs are played in the key of E major because the design of the guitar particularly lends itself to playing songs in that key. In this key the first A section is based on the E major chord, the second A section is based on the A major chord modulating back to the E major chord, and the B section starts with the B major chord, modulates to the A major chord, and then resolves to the E major chord. There are variations, but that is basically it.

This chord progression provides nice, comfortable, and predictable expectations of tonal variation for the listener. But this alone would be boring. Musical interest, variety, and tension are produced via the hexatonic "blues" scale melody line, where the third, fifth, and seventh notes of the scale are flattened to produce appealing dissonances over the repetitive chord progression. In addition, rhythmically the blues incorporates powerful hypnotic forward motion whose pleasant repetitive effect is commonly referred to as "the groove."

The blues started as a folk form but soon matured. As the now well-recognized standard musical features of the blues became established in the early twentieth century, blues composers began to publish their work. Later the first recordings were produced. The first published blues composition was likely "I Got the Blues," published in 1908 by New Orleans musician Antonio Maggio. Other early examples are Hart Wand's "Dallas Blues," published in 1912, and W. C. Handy's "The Memphis Blues," also published in 1912. Likely the first blues recording was "Crazy Blues," recorded in 1920 and sung by the African American vocalist Mamie Smith.

Early Life

Robert Johnson's humble origins and the fact that his musical brilliance wasn't recognized until decades after his death frustrates the study of his early life. The sketchy details of Johnson's chaotic childhood have been pieced together from a few diffi-cult-to-come-by facts and a lot of conjecture. Robert Johnson was most likely born on May 8, 1911, in Hazlehurst, Mississippi, to Julia Major Dodds and Noah Johnson. At the time, Julia Dodds and Noah Johnson were not married. Julia Dodds was instead married to Charles Dodds, a relatively well-to-do landowner and furniture maker. Shortly after Robert's birth, a lynch mob drove Charles Dodds from Hazlehurst to Memphis, Tennessee, where he changed his name to Charles Spencer. Not long after, Julia took Robert, now two years old, to live with her husband in Memphis.

In Memphis, Robert attended the Carnes Avenue Colored School, where he studied arithmetic, reading, language, music, and geography. At some point Julia Dodds left her husband, but Robert remained behind with Charles (Dodds) Spencer. At the age of ten, after living for several years in Memphis with Charles Spencer, Robert rejoined his mother, who had by that time mar-ried an illiterate sharecropper named Will "Dusty" Willis.

Later in life, Johnson's mother showed a preference for younger men. Although Julia Dodds was ten years younger than her husband Charles "Dodds" Spencer, Johnson's father was ten years younger than his mother, and her new husband was twenty-four years her junior. The newlyweds initially settled on a plan-tation in Lucas Township, Arkansas, but soon moved across the river to Commerce, Mississippi. Robert continued his education at the Indian Creek School in nearby Tunica, Mississippi, until he was at least sixteen years old. In later life, the ten or more years of public-school education he had received in Memphis and Tunica cast him as an intellectual within the contemporary blues music community, most of whom had little or no formal education at all.

When Robert was a teenager, his mother informed him that his real father was Noah Johnson. Robert then changed his name from Robert Spencer to Robert Johnson. By this time Robert Johnson had already established his longtime association with and love for the blues, likely starting to play the guitar when he was a young boy in Memphis. At the age of seventeen Robert Johnson married sixteen-year-old Virginia Travis. Shortly after their marriage, Virginia died in childbirth along with their child. Travis's family blamed Johnson for her death, saying it was a punishment for his singing secular songs and "selling his soul to the Devil." It is said that Johnson himself began to embrace the idea that his music might have come from an association with the devil. But it is not hard to imagine that the heartbreak of losing his young wife and child so soon after their marriage likely played a role in influencing him to spend the rest of his life singing sorrowful blues songs.

Traveling Blues Musician

Soon after the death of Johnson's wife, the blues musician Son House moved to nearby Robinsonville to join his musical partner, Willie Brown. Johnson's relationship with these musicians was a major episode in his development as a blues musician, although when interviewed by historians the two men remembered Johnson as a competent harmonica player but an embarrassingly bad guitarist.

Robert Johnson began his itinerant musical life leaving for Martinsville, Mississippi, to search for his natural father. In Martinsville he perfected his guitar style, learning from the master guitarist Isaiah "Ike" Zimmerman. Johnson lived with Zimmerman for a year, during which time Zimmerman taught him all he knew. They played together, touring local lumber camps and juke joints. Soon Johnson began performing on his own.

In 1930, while living in Martinsville, Mississippi, Johnson fathered a child after a very brief affair with a woman named Vergie Mae Smith. But for half a century no one seemed to know if the child had lived or what had happened to him or her. Johnson married Caletta Craft in May 1931 and in 1932 the couple lived for a while in Clarksdale, Mississippi. But this settled life did not last long, and Johnson soon left Clarksdale for a career as a "walking" or itinerant musician. Caletta died in early 1933.

From 1931 until his death in 1938, Robert Johnson lived the wandering musical life playing on street corners and in juke joints in small towns. He moved among Memphis, Tennessee; Helena, Arkansas; and small towns on the Mississippi Delta. He also on occasion traveled to distant destinations including Chicago, New York, Texas, Indiana, Kentucky, and Canada. In Indiana, Johnson played with the prominent blues guitarist Johnny Shines and in St. Louis with the blues musician Henry Townsend.

Robert Johnson formed many transient relationships on the road but also some long-term ones with women to whom he would return periodically. He based one of his best-known blues compositions, "Love in Vain," on his relationship with Willie Mae Powell, one of the women he knew during his touring days. But Johnson generally appeared to avoid intimacy. Most of the people whom he met on the road knew him superficially, with only sketchy knowledge of his life.

Everyone who met Johnson seemed to agree that he was well mannered and so soft-spoken that he was almost unintelligible. And while Johnson was pleasant and outgoing in public, in private he was reserved and independent. People remember him as a nice guy and in most ways fairly average, except for his musical talent and weaknesses for whiskey and women.

Whenever he arrived in a new town, Johnson would play for tips on street corners or in front of a local barbershop or restaurant. He had a gift for knowing which songs would most please his audiences, almost always not his own compositions or the music he enjoyed playing. Johnson was able to play a song after hearing it only once, so he quickly and easily accumulated a vast repertoire. He also had the performer's knack for establishing a rapport with his audiences. People would remember him months later if he visited again.

Johnson's appetite for whiskey and women ultimately led to his untimely demise. In 1938 Johnson was having an affair with a married woman named Beatrice Davis. Seeking revenge, her husband, Ralph, gave Johnson a bottle of whiskey into which he had dissolved mothballs, intending to sicken Johnson in order to teach him a lesson. But Johnson had developed a severe gastric ulcer after many years of heavy drinking. The mothballs irritated his ulcer, causing violent vomiting that led to gastrointestinal hemorrhage. Johnson bled to death on August 16, 1938, at the age of twenty-seven.

Recordings

In 1936 Johnson met Ernie Oertle, a salesman for the ARC group of record labels. Oertle then introduced Johnson to Don Law, a record producer who arranged a recording session for him in San Antonio, Texas, on November 23–25, 1936. Over the course of three days Johnson recorded sixteen compositions in an improvised, jury-rigged recording studio set up in a hotel room.

The songs he recorded included "Kind Hearted Woman Blues" (his first recording), "Come On in My Kitchen," "I Believe

I'll Dust My Broom," "Cross Roads Blues," "Terraplane Blues," "Last Fair Deal Gone Down," and "Sweet Home Chicago." A second recording session was held June 19–20, 1937, in Dallas, Texas, for Vitagraph (which became Warner Brothers records) using a similar primitive recording setup. During this second session Johnson recorded thirteen songs, eleven of which were released the following year. Johnson's lifetime discography from the two sessions comprised twenty-nine compositions.

"Terraplane Blues" and "Last Fair Deal Gone Down" were released shortly after the first session and were the only of his recordings Johnson would live to hear. Of these two, the more commercially successful was "Terraplane Blues," selling a modest five thousand copies. Later releases were "Milkcow's Calf Blues," "Love in Vain Blues," and "I Believe I'll Dust My Broom." None of these made much of an impact.

During the first half of the twentieth century, some early blues musicians achieved both success and recognition. Bessie Smith (1894–1937), known as the "Empress of the Blues," was a powerful vocalist and a successful businesswoman who sold tens of thousands if not hundreds of thousands of recordings. Blind Lemon Jefferson (1893–1929) achieved commercial success during the 1920s and is said to be the founding father of Texas blues. The blues guitarist Big Bill Broonzy (1893–1958) was one of the first musicians to bring the blues to Chicago, and he made a major impact in defining the city's sound. Leadbelly (1888–1949), born Huddie Ledbetter on a plantation near Mooringsport, Louisiana, had, despite multiple imprisonments, a long, commercially successful career that extended from the 1920s through the 1940s, including participation in the 1940s folk scene in New York City together with the folk music greats Woody Guthrie and Pete Seeger.

But Johnson suffered from the two classic problems that befall unrecognized geniuses. He was not in the right place at the right time, and very few people were exposed to his music. During Johnson's lifetime, the blues was a niche musical form. The audience was mainly African Americans, who generally lacked the money to pay for phonographs, blues records, or formal concerts. As an itinerant musician, Johnson played mostly in small towns in the South, almost never in large theaters or large musical venues. It was only in the mid-1930s and early 1940s, around the time of his death, that big band music, which is largely based on the blues, became the prominent popular American musical form. But it was too late for Johnson.

Rediscovery

In short, Johnson, an itinerant musician who played on street corners and in juke joints in small towns, performed a musical genre that did not enter the musical mainstream until after his death, and he died at the very young age of twenty-seven. While he was alive, he was hardly known or heard. According to Elijah Wald, an American folk blues guitarist and music historian, if someone had asked African American blues fans about Johnson during the first twenty years after his death, "the response in the vast majority of cases would have been a puzzled 'Robert who?'" Also according to Wald, "As far as the evolution of black music goes, Robert Johnson was an extremely minor figure, and very little that happened in the first decades following his death would have been affected if he had never played a note." But in 1961 Columbia Records released a compilation of Robert Johnson's recordings, the album *King of the Delta Blues Singers*, which finally exposed a wide, appreciative audience to Johnson's work.

Johnson was a master guitarist, one of the all-time greats on the instrument. His only guitar solo appears on his recording "Kind Hearted Woman Blues." To even highly accomplished players, this recording sounds like two guitarists are playing at the same time, one performing the guitar solo and the other providing accompaniment. When Keith Richards of the Rolling Stones first heard the solo, he asked his bandmate, Brian Jones, "Who is the other guy playing with him?" Richards recalled, "I was hearing two guitars, and it took a long time to actually realize he was doing it all by himself." Richards also said that Johnson "was like an orchestra all by himself."

For his recording of "I Believe I'll Dust My Broom," Johnson fashioned a boogie style bass line that had previously only been performed on the piano. Reggie Ugwu wrote in the *New York Times* that Johnson, "imitating the boogie-woogie style of piano playing, used his guitar to play rhythm, bass, and slide simultaneously, all while singing." This was completely new at the time, although thanks to Johnson this technique eventually became a standard part of the blues guitar style.

Contemporary music critics have rated Johnson as one of the greatest guitarists of all time based on just this handful of old recordings. In 1990, on the fifty-second anniversary of his death, *Spin* rated Johnson first on its list of "35 Guitar Gods." In 2008, seventy years after he died, *Rolling Stone* ranked Johnson fifth on their list of "100 Greatest Guitarists of All Time." And in 2010, seventy-two years after his death, Guitar.com ranked him ninth in its list of "Top 50 Guitarists of All Time." Professional guitarists who cited Johnson as a major influence include Keith Richards, Jimi Hendrix, and Eric Clapton.

Johnson was also extremely influential as a composer and

blues stylist. If you listen today to Johnson's first recording, "Kind Hearted Woman Blues," it sounds as if he had listened to one hundred of the top contemporary blues records and composed a song that distilled the essence and the best features that all those songs held in common into a single composition. But the reverse is true. The composers of the current top one hundred blues recordings listened to Johnson's work early in their careers and based their writings on the forms Johnson created.

One of the featured compositions in the 1980 movie *Blues Brothers* is "Sweet Home Chicago," a number played by the Blues Brothers band and sung by John Belushi and Dan Aykroyd. In the movie, the song sounds fresh and dynamic, and given its contemporary feel and the fact that the movie is set in Chicago, it would be easy to think that the producers had hired a composer to write the song specifically for the film. Wrong. "Sweet Home Chicago" was one of the compositions Robert Johnson recorded in 1936. The movie version and many other contemporary versions, including one by Eric Clapton, are covers of the Robert Johnson original.

Over the years the Rolling Stones recorded three Robert Johnson songs. Mick Jagger performed excerpts of two Robert Johnson songs for the movie *Performance*. One of the best-known of these is the Stone's recording of "Love in Vain," Johnson's sadly bleak song about unrequited love, recorded by him in 1937 during his last recording session and issued as the last of his original 78 rpm records.

Eric Clapton considers Johnson to be "the most important blues musician who ever lived." Clapton recorded an entire album of Robert Johnson's songs, *Me and Mr. Johnson*, released in 2004 as a tribute to the inspiring bluesman. The song "Crossroads," recorded with Cream in 1968, was their arrangement of Johnson's "Cross Road Blues." It has been said that Clapton was the person responsible for making Robert Johnson a household name among rock musicians.

Robert Plant of Led Zeppelin said Robert Johnson was someone "to whom we all owed our existence, in some way." Led Zeppelin recorded their own version of the Robert Johnson song "Traveling Riverside Blues." Fleetwood Mac was also strongly influenced by Johnson. Fleetwood Mac cofounder Jeremy Spencer recorded two Robert Johnson covers while his fellow cofounder Peter Green recorded all of Johnson's work in two albums, *The Robert Johnson Songbook* and *Hot Foot Powder.*

Bob Dylan wrote of Johnson in his 2004 autobiography *Chronicles: Volume One*, "If I hadn't heard the Robert Johnson record when I did, there probably would have been hundreds of lines of mine that would have been shut down—that I wouldn't have felt free enough or upraised enough to write."

Financial Legacy

Johnson died penniless, but the rediscovery of his music in 1961 and its subsequent adoption by famous rock musicians soon produced a steady if small stream of royalties. Blues aficionado Stephen LaVere started managing the Robert Johnson songbook in 1974 and agreed to split the royalties with Carrie Thompson, Robert Johnson's half-sister, who had filed as Robert Johnson's next of kin and was thus awarded by the probate court a share in the royalties from Johnson's compositions.

In 1983, Carrie died, leaving her stepsister Annye Anderson, who was not related to Robert Johnson, to manage her affairs, and in 1989, Annye became the administrator of both Carrie's and Robert Johnson's estates. No one seemed to care much about it because the royalty stream from Johnson's compositions remained extremely modest. But in 1990 Columbia Records released the album *Robert L. Johnson: The Complete Recordings.* The album won a Grammy Award, sold over 500,000 copies, and was a huge financial success, generating millions of dollars in royalties. Suddenly, dozens of people stepped forward to claim they were descendants of Robert Johnson. All were deemed frauds and obvious impostors until one day Claud L. Johnson appeared before the court. He claimed that he was the son of Robert Johnson and Virginia Smith, born nine months after their brief (some say one-night) affair in 1930.

Mack McCormick was a Texas cultural historian studying the life of Robert Johnson who, as part of his research, had traveled to Crystal Springs, Mississippi, in 1970. There he met an old woman named Vergie Mae Smith who claimed that Robert Johnson was the father of her son Claud. The results of McCormick's scholarship eventually made its way to Stephen LaVere, the manager of Johnson's estate.

Not long thereafter Claud Johnson received a summons to appear before the probate court. He had no idea what to do and sought advice from Jim Kitchens, a Jackson, Mississippi, trial lawyer and former district attorney who was a regular customer at Claud Johnson's barbeque restaurant in Crystal Springs. After hearing his story, Kitchens agreed to represent him when he appeared before the court.

In 1992, long before the advent of routine DNA testing, Claud Johnson's claim that he was Robert Johnson's son turned on the salacious testimony of one Eula Mae Williams. The white-haired octogenarian stated under oath that while she and her boyfriend were having sexual intercourse, she had observed Robert Johnson and Vergie Mae Smith having sexual intercourse very close by at exactly the same time. Under questioning she insisted that she had not been so distracted by her own activities as to be in any way uncertain regarding what Robert Johnson and Vergie Mae Smith had been doing. And after their brief encounter Claud was born nine months later.

In the stew of multiple paternity claims and counterclaims, the case wound its way through the courts for eight years, including two trips to the Mississippi Supreme Court and two to the United States Supreme Court. In the interim, Vergie Mae Smith, Claud's mother, died, but in the end a whopping sixty-two years after Robert Johnson's death, his estate was finally settled and Claud Johnson was determined to be his sole heir. After the judgment, Claud's lawyer promptly handed him a six-figure cashier's check, more money than Claud Johnson had seen in his entire life.

Claud Johnson had been raised by his grandfather, a strict Baptist preacher who had no use for the devil's blues music. Only gospel music was permitted in the house. In addition to his taste in music, Claud Johnson was unlike his father in almost every imaginable way. Claud was a dedicated family man. He and his wife, Earnestine, were married for sixty years and, despite living a hardscrabble life and having almost no money, sent all six of their children to college.

Claud was an extraordinarily diligent, hard worker. He started working long hours at physically demanding jobs right after dropping out of school in the sixth grade, and he sometimes worked two or three jobs at the same time. Over the years, he worked unloading trucks, in a saw mill, in a bottling plant, in a service station, in an electrical power plant, and for a cabinet maker. He owned a barbeque restaurant in Crystal Springs, Mississippi, and at one point bought a dilapidated Mack gravel truck to supplement his income by working as an independent driver hauling sand and gravel for Green Brothers Gravel Company in Crystal Springs.

After many, many years of hard labor in low-paying jobs, the inheritance from Robert Johnson's estate was, to say the least, life-changing for Claud Johnson. He purchased a gated estate on forty-seven acres of land with the money. A long, curving driveway leads from the pink brick gate on the main road to Claud's mansion. Inside there was more space than Claud ever could have imagined he would own. The second floor was mostly an empty expanse of cream-colored wall-to-wall carpeting. In 2004 Claud told a *Los Angeles Times* reporter that he had so much space on the first floor he considered the second story superfluous. He said he wasn't sure his wife was ever up there.

But Claud never forgot his life before the inheritance. He remained sentimental about the broken-down Mack gravel truck that for years had provided him with valuable supplementary income, and he parked the truck next to the garage of his magnificent estate, where a different owner might have parked his prized Ferrari sports car. Like the hardworking man he always was, Claud Johnson still mowed his own lawn. Although Robert Johnson never lived long enough to see his music become recognized and appreciated, his son Claud, Claud's wife, and his six grandchildren were able to enjoy the fruits of Robert Johnson's creative genius.

Suggested Readings

Chapter 1: Max Planck

Bembenek, Scott. "Einstein and the Quantum." *Observations* (blog). *Scientific American*, March 27, 2018. https://blogs.scientificamerican.com/observations/einstein-and-the-quantum/.

Brown, Brandon R. *Planck: Driven by Vision, Broken by War.* Oxford: Oxford University Press, 2015.

Charles Rivers Editors. *Max Planck: The Life and Legacy of the Influential German Physicist Who Pioneered Quantum Theory.* Charles Rivers Editors, 2018.

Einstein, Albert, Max Planck, and W. Heisenberg. *Colleagues in Genius: Out of My Later Years, Scientific Autobiography, and Nuclear Physics.* New York: Philosophical Library/Open Road, 2019.

Fergus, Julie. "Max Planck and Albert Einstein." *OUPblog.* Oxford University Press, November 23, 2015. https://blog.oup.com/2015/11/max-planck-albert-einstein/.

Heilbron, J. L. *The Dilemmas of an Upright Man: Max Planck and the Fortunes of German Science, With a New Afterword.* Rev. ed. Cambridge: Harvard University Press, 2000.

Kiger, Patrick J. "What Is Planck's Constant, and Why Does the Universe Depend on It?" How Stuff Works. Updated June 9, 2023. https://science.howstuffworks.com/dictionary/physics-terms/plancks-constant.htm.

Kuhn, Thomas S. *The Structure of Scientific Revolutions: 50th Anniversary Edition*. 4th ed. Chicago: University of Chicago Press, 2012.

Kwai, Isabella, Cora Engelbrecht, and Dennis Overbye. "Nobel Prize in Physics Is Awarded to Three Scientists for Work Exploring Quantum Weirdness." *New York Times*, October 4, 2022. https://www.nytimes.com/2022/10/04/science/nobel-prize-physics-winner.html.

"Max Planck: the Reluctant Revolutionary." Physicsworld, December 1, 2000. https://physicsworld.com/a/max-planck-the-reluctant-revolutionary/.

Planck, Max. *Scientific Autobiography: And Other Papers*. New York: Philosophical Library/Open Road, 2014.

Rosenberg, Jennifer. "Max Planck Formulates Quantum Theory." ThoughtCo. Updated April 15, 2018. https://www.thoughtco.com/max-planck-formulates-quantum-theory-1779191.

Stein, James. "Planck's Constant: The Number That Rules Technology, Reality, and Life." NOVA, October 24, 2011. https://www.pbs.org/wgbh/nova/article/plancks-constant/.

Chapter 2: Gregor Mendel

Fairbanks, Daniel J. *Gregor Mendel: His Life and Legacy*. Lanham, MD: Prometheus, 2022.

Henig, Robin Marantz. *The Monk in the Garden: The Lost and Found Genius of Gregor Mendel, the Father of Genetics*. Boston: Mariner Books, 2017.

Mawer, Simon. *Gregor Mendel: Planting the Seeds of Genetics*. New York: Abrams Books, 2006.

Olby, Robert C. *Origins of Mendelism*. New York: Schocken Books, 1966.

Pincas, Christina. "Gregor Mendel and the Art of Mis-communication." Scientopia, August 2, 2012. https://guestblog.scientopia.org/2012/08/02/gregor-mendel-and-the-art-of-mis-communication/.

Pincas, Christina. "Mud Sticks, Especially If You Are Gregor Mendel." Scientopia, August 3, 2012. https://guestblog.scientopia.org/2012/08/03/mud-sticks-especially-if-you-are-gregor-mendel/.

Reinberger, H-J. "When Did Carl Correns Read Gregor Mendel's Paper?" *Isis* 86 (1995): 612–616.

Chapter 3: Barbara McClintock

Comfort, Nathaniel C. *The Tangled Field: Barbara McClintock's Search for the Patterns of Genetic Control*. Cambridge, MA: Harvard University Press, 2001.

Keller, Evelyn Fox. *A Feeling for the Organism, 10th Anniversary Edition: The Life and Work of Barbara McClintock*. Chicago: Times Books, 1984.

Kidwell, Margaret G., and Damon R. Lisch. "Perspective: Transposable Elements, Parasitic DNA and Genome Evolution." *Evolution* 55 (2001): 1–24.

Kirschner, Marc. "Bruce Alberts, *Science*'s New Editor." *Science* 319 (February 29, 2008): 1199.

Miko, Ilona. "Thomas Hunt Morgan and Sex Linkage." *Nature Education* 1 (2008): 143. https://www.nature.com/scitable/topicpage/thomas-hunt-morgan-and-sex-linkage-452/.

Pinsker, Wilhelm, et al. "The Evolutionary Life History of P Transposons: From Horizontal Invaders to Domesticated Neogenes." *Chromosoma* 110 (2001): 148–158.

Ravindran, Sandeep. "Barbara McClintock and the Discovery of Jumping Genes." *Proceedings National Academy Science* 109 (2012): 20198–20199.

Sarchet, Penny. "Barbara McClintock." *New Scientist*. Accessed October 18, 2022. https://www.newscientist.com/people/barbara-mcclintock/.

Chapter 4: Galileo Galilei

Burns, Joseph A. "The Four Hundred Years of Planetary Science since Galileo and Kepler." *Nature* 466 (2010): 575–584.

Galilei, Galileo. *Dialogues Concerning Two New Sciences*. Overland Park, KS: Digireads.com, 2011.

Galilei, Galileo. *The Essential Galileo*. Translated by Maurice A. Finocchiaro. Indianapolis: Hackett Classics, 2008.

Galilei, Galileo. *Sidereus Nuncius, or The Sidereal Messenger*. Translated by Albert Van Helden. Chicago: University of Chicago Press, 2016.

"Galileo Galilei (1564–1642)." *British Journal of Sports Medicine* 40 (September 2006): 806–807.

Jack, Albert. *They Laughed at Galileo: How the Great Inventors Proved Their Critics Wrong*. New York: Skyhorse, 2015.

Pokorny, Martin. "Translation of Jan Patočka's 'Galileo Galilei and the End of the Ancient Cosmos.'" *Studies in Eastern European Thought* 73 (2021): 367–375.

Chapter 5: Ignaz Semmelweis

Ataman, Ahmet Doğan, Emine Elif Vatanoğlu-Lutz, and Gazi Yıldırım. "Medicine in Stamps-Ignaz Semmelweis and Puerperal Fever." *Journal of the Turkish German Gynecological Association* 14 (2013): 35–39.

Carter, K. Codell, and Barbara R. Carter. *Childbed Fever: A Scientific Biography of Ignaz Semmelweis*. London: Routledge Books, 2017.

Cavaillon, Jean-Marc, and Fabrice Chrétien. "From Septicemia to Sepsis 3.0—From Ignaz Semmelweis to Louis Pasteur." *Genes and Immunity* 20 (2019): 371–382.

Gupta, Vipin K., et al. "Semmelweis Reflex: An Age-Old Prejudice." *World Neurosurgery*, 136 (2020): e116–e125.

Kandar, Nicholas. "Ignaz Semmelweis: The Savior of Mothers; On the 200th Anniversary of the Birth." *American Journal of Obstetrical Gynecology* 219 (2018): 519–522.

Semmelweis, Ignaz. *Etiology, Concept, and Prophylaxis of Childbed Fever*. Translated by K. Codell Carter. Madison: University of Wisconsin Press, 1983.

Vermeil, T., et al. "Hand Hygiene in Hospitals: Anatomy of a Revolution." *Journal of Hospital Infection* 101 (2019): 383–392.

Chapter 6: Peyton Rous

Andrews, C. H. "Francis Peyton Rous." *Biographical Memoirs of Fellows of the Royal Society* 17 (1971): 643–662.

DeVita, Vincent T., and Edward Chu. "A History of Cancer Chemotherapy." *Cancer Research* 68 (2008): 8643–8653.

Dobosz, Paula, and Tomasz Dzieciatkowski. "The Intriguing History of Cancer Immunotherapy." *Frontiers in Immunology* 10 (2019): 1–10.

Hajdu, Steven I. "A Note From History: Landmarks in History of Cancer, Part 1." *Cancer* 117 (March 1, 2011): 1097–1102.

Hajdu, Steven I., and Farbod Darvishian. "A Note From History: Landmarks in History of Cancer, Part 5." *Cancer* 119 (April 15, 2013): 1450–1466.

Lipsick, Joseph. "A History of Cancer Research: Carcinogens and Mutagens." *Cold Spring Harbor Perspectives on Medicine* 11 (2021): a035857.

Lipsick, Joseph. "A History of Cancer Research: Tumor Viruses." *Cold Spring Harbor Perspectives on Biology* 13 (2021): a035774.

Luo, Ji. "Principles of Cancer Therapy: Oncogene and Non-oncogene Addiction." *Cell* 136 (2009): 823–837.

Miller, G., and J. Stebbing. "Thirty Years of Oncogene." *Oncogene* 37 (2018): 553–554.

Newman, Barclay Moon. "Cancer's Mysterious Puzzle." *Scientific American* 161 (1939): 278–280.

"Peyton Rous – Biographical." The Nobel Prize (website). Accessed October 21, 2022. https://www.nobelprize.org/prizes/medicine/1966/rous/biographical/.

Pomeroy, R. "Disproved Discoveries That Won Nobel Prizes." Real Clear Science, October 7, 2015. https://www.realclearscience.com/blog/2015/10/nobel_prizes_awarded_for_disproved_discoveries.html.

Stehelin, D., et al. "DNA Related to the Transforming Gene(s) of Avian Sarcoma Viruses Is Present in Normal Avian DNA." *Nature* 260 (1976): 170–173.

Steven, Martin G. "The Hunting of the Src." *Nature Reviews Molecular Cellular Biology* 2 (2001): 467–475.

Tontonoz, Matthew. "How a Chicken Helped Solve the Mystery of Cancer." Memorial Sloan Kettering Cancer Center, December 27, 2017. https://www.mskcc.org/news/how-chicken-helped-solve-mystery.

Van Epps, Heather L. "Peyton Rous: Father of the Tumor Virus." *Journal of Experimental Medicine* 201 (2005): 320.

Weiss, Robin A., and Peter K. Vogt. "100 Years of Rous Sarcoma Virus." *Journal of Experimental Medicine* 208 (2011): 2351–2355.

Zheng, Li, et al. "Lessons Learned from BRCA1 and BRCA2." *Oncogene* 19 (2000): 6159–6175.

Chapter 7: Roger Revelle

Cohen, Jon. "Was Underwater 'Shot' Harmful to the Whales?" *Science* 252 (1991): 912–914.

Drexler, Madeline. "Population Visionary." T. H. Chan School of Public Health, Harvard, Fall 2013. https://www.hsph.harvard.edu/news/magazine/centennial-population-visionary/.

Frieman, Edward A. "Roger Randall Dougan Revelle." Revelle College, UC San Diego website. Accessed October 21, 2022. https://revelle.ucsd.edu/about/roger-revelle.html.

Morgan, Judith, and Neil Morgan. *Roger: A Biography of Roger Revelle.* La Jolla CA: Scripps Institution of Oceanography, 1996.

Munk, Walter H. "Tribute to Roger Revelle and His Contribution to Studies of Carbon Dioxide and Climate Change." *Proceedings National Academy Science* 94 (1997): 8275–8279.

Revelle, Roger. "Munk's Experiment." *Science* 153 (1991): 118.

Riebeek, Holli. "Paleoclimatology: Introduction." NASA Earth Observatory, June 28, 2005. https://earthobservatory.nasa.gov/features/Paleoclimatology/paleoclimatology_intro.php.

Ringrose, Kathryn. "Interview with Dr. Roger Revelle." University of California, San Diego 25th Anniversary Oral History Project. May 15–16, 1985. https://library.ucsd.edu/speccoll/siooralhistories/2010-44-Revelle.pdf.

Tans, P. P., et al. "Natural Atmospheric ^{14}C Variation and the Suess Effect." *Nature* 280 (1979): 826–827.

Weart, Spencer R. *The Discovery of Global Warming: Revised and Expanded Edition*. Cambridge, MA: Harvard University Press, 2008.

Chapter 8: Rachel Carson

Axelson, Gustave. "Nearly 30% of Birds in U.S., Canada Have Vanished since 1970." *Cornell Chronicle*, September 19, 2019. https://news.cornell.edu/stories/2019/09/nearly-30-birds-us-canada-have-vanished-1970.

Brinkley, Douglas. *Silent Spring Revolution: John F. Kennedy, Rachel Carson, Lyndon Johnson, Richard Nixon, and the Great Environmental Awakening*. New York: Harper, 2022.

Carson, Rachel. *Rachel Carson: The Sea Trilogy: Under the Sea-Wind / The Sea Around Us / The Edge of the Sea*. New York: Library of America, 2021.

Carson, Rachel. *Silent Spring Anniversary Edition*. Boston: Mariner Books, 2022.

Lear, Linda. *Rachel Carson: Witness for Nature*. Boston: Mariner Books, 2009.

The Life and Legacy of Rachel Carson (website). Accessed October 22, 2022. http://www.rachelcarson.org.

Light, John. "11 Ways the EPA Has Helped Americans." BillMoyers.com, March 17, 2017. https://billmoyers.com/story/11-ways-epa-helped-americans/.

Pappas, Stephanie. "Irish Potato Blight Originated in South America." *LiveScience*, January 3, 2017. https://www.livescience.com/57363-irish-potato-blight-originated-in-south-america.html.

Russell, Edmund. *War and Nature: Fighting Humans and Insects with Chemicals from World War I to Silent Spring*. Cambridge: Cambridge University Press, 2001.

Zimmer, Carl. "E. O. Wilson, a Pioneer of Evolutionary Biology, Dies at 92." *New York Times*, December 27, 2021. https://www.nytimes.com/2021/12/27/science/eo-wilson-dead.html.

Chapter 9: Stanley Prusiner

Aguzzi, Adriano, et al. "Molecular Mechanisms of Prion Pathogenesis." *Annual Reviews of Pathology* 3 (2008): 11–40.

"Amedeo Avogadro." Famous Scientists, January 12, 2015. https://www.famousscientists.org/amedeo-avogadro/.

Brown, Paul, and Raymond Bradley. "1755 and All That: A Historical Primer of Transmissible Spongiform Encephalopathy." *British Medical Journal* 317 (1998): 19–26.

Khot, Sandeep, et al. "The Vietnam War and Medical Research: Untold Legacy of the U.S. Doctor Draft and the NIH 'Yellow Berets.'" *Academic Medicine* 86 (2011): 502–508.

Liberski, Pawel P., et al. "Kuru, the First Human Prion Disease." *Viruses* 11 (2019): 232–257.

Lindquist, Susan. "Mad Cows Meet Psi-chotic Yeast: The Expansion of the Prion Hypothesis." *Cell* 89 (1997): 495–498.

Nathanson, Neal, et al. "Bovine Spongiform Encephalopathy (BSE): Causes and Consequences of a Common Source Epidemic." *American Journal of Epidemiology* 145 (1997): 959–969.

Prusiner, Stanley B. "Novel Proteinaceous Infectious Particles Cause Scrapie." *Science* 216 (1982): 136–144.

Schneider, Kurt, et al. "The Early History of the Transmissible Spongiform Encephalopathies Exemplified by Scrapie." *Brain Research Bulletin* 77 (2008): 343–355.

"Stanley B. Prusiner – Biographical." The Nobel Prize (website). Accessed October 24, 2022. https://www.nobelprize.org/prizes/medicine/1997/prusiner/biographical/.

Zabel, Mark D., and Crystal Reid. "A Brief History of Prions." *FEMS Pathogens and Disease* 73 (2015): 1–8.

Chapter 10: Amedeo Avogadro

Aglio, Linda S., et al. "History of Anaesthesia: Amedeo Avogadro (1776–1856)—Do His Accomplishments Match his Reputation?" *European Journal of Anesthesiology* 33 (2016): 1–3.

"Amedeo Avogadro." Science History Institute. Accessed October 24, 2022. https://www.sciencehistory.org/historical-profile/amedeo-avogadro.

Constable, Edwin C., et al. "John Dalton: The Man and the Myth." *Dalton Transactions*, 51 (2022): 768–776.

Hinshelwood, Cyril N. "Amedeo Avogadro." *Science* 124 (1956): 708–710.

Newburgh, Ronald, et al. "Einstein, Perrin, and the Reality of Atoms: 1905 Revisited." *American Journal of Physics* 74 (2006): 478–481.

Pauling, Linus. "Amedeo Avogadro." *Science* 124 (1956): 710–713.

Pyle, Andrew. "Atoms and Atomism." In *The Classical Tradition*, edited by Anthony Grafton, Glenn W. Most, and Salvatore Settis, 103–104. Cambridge MA: Belknap Press of Harvard University Press, 2010.

Chapter 11: David Cushman and Miguel Ondetti

"ACE inhibitors for treating hypertension." Lasker Foundation website. Accessed October 24, 2022. https://laskerfoundation.org/winners/ace-inhibitors-for-treating-hypertension/.

Black, Henry R. "Antihypertensive Therapy and Cardiovascular Disease Impact of Effective Therapy on Disease Progression." *American Journal of Hypertension* 11 (1998): 3S–8S.

Cushman, D. W., H. S. Cheung, E. F. Sabo, and M. A. Ondetti. "Design of New Antihypertensive Drugs: Potent and Specific Inhibitors of Angiotensin-converting Enzyme." *Progress in Cardiovascular Disease* 21 (1978): 176–182.

Cushman D. W., H. S. Cheung, E. F. Sabo, and M. A. Ondetti. "Design of Potent Competitive Inhibitors of Angiotensin-Converting Enzyme. Carboxyalkanoyl and Mercaptoalkanoyl Amino Acids." *Biochemistry* 16 (1977): 5484–5491.

Cushman, D. W., and M. A. Ondetti. "Design of Angiotensin Converting Enzyme Inhibitors." *Nature Medicine* 5 (1989): 1110–1112.

Cushman, D. W., and M. A. Ondetti. "History of the Design of Captopril and Related Inhibitors of Angiotensin Converting Enzyme." *Hypertension* 17 (1991): 589–592.

Cushman, D. W., and M. A. Ondetti. "Inhibitors of Angiotensin-converting Enzyme for Treatment of Hypertension." *Biochemical Pharmacology* 29 (1980): 1871–1877.

Franklin, Stanley S., and Nathan D. Wong. "Hypertension and Cardiovascular Disease: Contributions of the Framingham Heart Study." *Global Heart* 8 (2013): 49–57.

Ondetti, M. A., B. Rubin, and D. W. Cushman. "Design of Specific Inhibitors of Angiotensin-Converting Enzyme: New Class of Orally Active Antihypertensive Agents." *Science* 196 (1977): 441–444.

Opie, Lionel H., and Helmut Kowolik. "The Discovery of Captopril: From Large Animals to Small Molecules." *Cardiovascular Research* 30 (1995): 18–25.

Rubin, Ronald P. "Sir John Robert Vane and the Mode of Action of Aspirin." *International Journal of Research Studies in Medical and Health Sciences* 5 (2020): 38–42.

Welch, Arnold D. "Reminiscences in Pharmacology: Auld Acquaintance Ne'er Forgot." *Annual Review Pharmacology and Toxicology* 25 (1985): 1–26.

Chapter 12: Surendra Nath "Suren" Sehgal

Boyd, Ashleigh S. *Immunological Tolerance: Methods and Protocols.* Totowa, NJ: Humana, 2019.

Dangoor, Joseph Yoav, et al. "Transplantation: A Brief History." *Experimental and Clinical. Transplantation* 1 (2015): 1–5.

Linden, Peter K. "History of Solid Organ Transplantation and Organ Donation." *Critical Care Clinics* 25 (2009): 165–184.

Livi, George P. "Halcyon Days of TOR: Reflections on the Multiple Independent Discovery of the Yeast and Mammalian TOR Proteins." *Gene* 692 (2019): 145–155.

Loria, Kevin. "A Rogue Doctor Saved a Potential Miracle Drug by Storing Samples in His Home after Being Told to Throw Them Away." *Insider*, February 20, 2015. https://www.businessinsider.com/suren-sehgal-saved-rapamycin-anti-aging-drug-2015-2.

Morris, Peter J. "Transplantation: A Medical Miracle of the 20th Century." *New England Journal of Medicine* 351 (2004): 2678–2680.

"Peter Medawar – Biographical." The Nobel Prize (website). Accessed October 24, 2022. https://www.nobelprize.org/prizes/medicine/1960/medawar/biographical/.

Powell, Jonathan D., et al. "Regulation of Immune Responses by mTOR." *Annual Review of Immunology* 30 (2012): 39–68.

Samanta, Debopam. "Surendra Nath Sehgal: A Pioneer in Rapamycin Discovery." *Indian Journal of Cancer* 54 (2017): 697–699.

Sayegh, Mohamed H., and Charles B. Carpenter. "Transplantation 50 Years Later—Progress, Challenges, and Promises." *New England Journal of Medicine* 351 (2004): 2761–2766.

Schlich, Thomas. "The Art of Medicine: The Origins of Organ Transplantation." *Lancet* 378 (2004): 1372–1373.

Sehgal, Ajai. "Suren Sehgal." 2003. http://www.sehgal.net/surenshistory.htm.

Shayan, Hossein. "Organ Transplantation: From Myth to Reality." *Journal of Investigative Surgery* 14 (2001): 135–138.

Simpson, Elizabeth. "Medawar's Legacy to Cellular Immunology and Clinical Transplantation: A Commentary on Billingham, Brent and Medawar

(1956) 'Quantitative Studies on Tissue Transplantation Immunity. III. Actively Acquired Tolerance.'" *Philosophical Transactions* B 370 (2014): 1–12.

Slater, Stephan. "The Discovery of Thyroid Replacement Therapy. Part 3: A Complete Transformation." *Journal of the Royal Society of Medicine* 104 (2010): 100–106.

Toledo-Pereyra, Luis H. "Nobel Laureate Surgeons." *Journal of Investigative Surgery* 19 (2006): 211–218.

Chapter 13: Alfred Wegener

Ball, Philip. "Lessons from Cold Fusion, 30 Years On." *Nature* 569 (2019): 601.

Doménech, Francisco. "Wegener and His Theory of Continental Drift That Broke with Geologists." OpenMind BBVA, October 30, 2020. https://www.bbvaopenmind.com/en/science/leading-figures/alfred-wegener-theory-of-continental-drift/.

Eiseley, Loren C. "Alfred Russel Wallace." *Scientific American* 200 (1959): 70–84.

Frankel, Henry R. *The Continental Drift Controversy*. Cambridge: Cambridge University Press, 2012.

Frisch, Wolfgang, et al. *Plate Tectonics: Continental Drift and Mountain Building*. Berlin: Springer, 2010.

Morgan, W. Jason. "Rises, Trenches, Great Faults, and Crustal Block." *Journal of Geophysical Research* 73 (1968): 1959–1981.

Oskin, Becky. "Continental Drift: The Groundbreaking Theory of Moving Continents." Live Science, December 14, 2021. https://www.livescience.com/37529-continental-drift.html.

Rogers, John J. W., and M. Santosh. *Continents and Supercontinents*. Oxford: Oxford University Press, 2004.

Romano, Marco, and Francesco Latino Chiocci. "Celebrating Marie Tharp." *Science* 370 (2020): 1415.

Wegener, Alfred. *The Origin of Continents and Oceans*. Mineola, NY: Dover Publications, 2011.

Wolchover, Natalie, et al. "Jason Morgan Recalls Discovering Earth's Tectonic Plates." *Quanta Magazine*, August 28, 2017. https://www.quantamagazine.org/jason-morgan-recalls-discovering-earths-tectonic-plates-20170828/.

Chapter 14: Robin Warren and Barry Marshall

Alkim, Huseyin, et al. "Role of Bismuth in the Eradication of *Helicobacter pylori*." *American Journal of Therapeutics* 24 (2017): e751–e757.

Graham, David Y. "History of *Helicobacter pylori*, Duodenal Ulcer, Gastric Ulcer, and Gastric Cancer." *World Journal of Gastroenterology* 20 (2014): 5191–5204.

Hanley, Brian P., et al. "Review of Scientific Self-Experimentation: Ethics History, Regulation, Scenarios, and Views Among Ethics Committees and Prominent Scientists." *Rejuvenation Research* 22 (2019): 31–42.

Marshall, Barry J. "Nobel Lecture." The Nobel Prize (website). Accessed October 27, 2022. https://www.nobelprize.org/prizes/medicine/2005/marshall/lecture/.

Marshall, Barry J., and Paul C. Adams. "*Helicobacter pylori*: A Nobel Pursuit?" *Canadian Journal of Gastroenterology* 22 (2008): 895–896.

Marshall, Barry J., and J. Robin Warren. "Unidentified Curved Bacilli in the Stomach of Patients with Gastritis and Peptic Ulceration." *Lancet* 1, no. 8390 (June 16, 1984): 1311–1315.

Sivak, M. V. "Gastrointestinal Endoscopy: Past and Future." *Gut* 55 (2006): 1061–1064.

Sung, Joseph Y. J. "Marshall and Warren Lecture 2009: Peptic Ulcer Bleeding: An Expedition of 20 Years from 1989–2009." *Journal of Gastroenterology and Hepatology* 25 (2010): 229–233.

Warren, J. Robin. "Nobel Lecture." The Nobel Prize (website). Accessed

October 27, 2022. https://www.nobelprize.org/prizes/medicine/2005/warren/lecture/.

Warren, J. Robin, and B. Marshall. "Unidentified Curved Bacilli on Gastric Epithelium in Active Chronic Gastritis." *Lancet* 1, no. 8336 (June 4, 1983): 1273–1275.

Epilogue

Appasani, Krishnarao. *Genome Editing and Engineering: From TALENs, ZFNs, and CRISPRs to Molecular Surgery.* Cambridge: Cambridge University Press, 2017.

Latour, B., and Woolgar, S. *Laboratory Life: The Construction of Scientific Facts.* 2nd ed. Princeton, NJ: Princeton University Press, 1986.

Marks, Lara V. *The Lock and Key of Medicine: Monoclonal Antibodies and the Transformation of Healthcare.* New Haven, CT: Yale University Press, 2015.

Robinow, Paul. *Making PCR: A Story of Biotechnology.* Chicago: University of Chicago Press, 1997.

Appendix

Anderson, Annye C. *Brother Robert: Growing Up with Robert Johnson.* New York: Hachette Books, 2020.

Barry, Ellen. "Bluesman's Son Gets His Due." *Los Angeles Times,* June 2, 2004. https://www.latimes.com/archives/la-xpm-2004-jun-02-na-claud2-story.html.

"Bluesman's Estate Finally Settles 62 Years After Death." SaveWealth.com. Accessed October 17, 2022. http://www.savewealth.com/news/0006/johnson/.

"Claud L. Johnson (December 16, 1931 – June 30, 2015)." Robert Johnson Blues Foundation website. July 1, 2015. https://www.robertjohnsonbluesfoundation.org/news/claud-l-johnson-december-16-1931-june-30-2015/.

Graves, Tom. *Crossroads*. Self-published, BookBaby, 2008.

Greenburg, Alan. *Love in Vain: A Vision of Robert Johnson*. Minneapolis: University of Minnesota Press, 2012.

Guralnick, Peter. *Searching for Robert Johnson: The Life and Legend of the "King of the Delta Blues Singers."* Boston: Little, Brown, 2020.

Komura, Edward. *The Road to Robert Johnson: The Genesis and Evolution of Blues in the Delta from the Late 1800s through 1938*. Milwaukee: Hal Leonard Publishing, 2007.

Ugwu, Reggie. "Overlooked No More: Robert Johnson, Bluesman Whose Life Was a Riddle." *New York Times*, September 25, 2019. https://www.nytimes.com/2019/09/25/obituaries/robert-johnson-overlooked.html.

Wald, Elijah. *Escaping the Delta: Robert Johnson and the Invention of the Blues*. New York: Amistad, 2004.

Index